Nature Conservation in the New Economy

Nature Conservation in the New Economy

People, Wildlife and the Law in India

edited by

Ghazala Shahabuddin

K. Sivaramakrishnan

Orient BlackSwan

NATURE CONSERVATION IN THE NEW ECONOMY:
PEOPLE, WILDLIFE AND THE LAW IN INDIA

ORIENT BLACKSWAN PRIVATE LIMITED

Registered Office
3-6-752 Himayatnagar, Hyderabad 500 029, Telangana, India
e-mail: centraloffice@orientblackswan.com

Other Offices
Bengaluru, Bhopal, Chennai, Guwahati, Hyderabad, Jaipur, Kolkata, Lucknow, Mumbai, New Delhi, Noida, Patna, Visakhapatnam

First published 2019

ISBN 978-93-5287-613-6

029607

Typeset in ITC Baskerville Std 10/12 by
Jojy Philip, New Delhi 110 015

Printed in India at
Glorious Printers, Delhi

Published by
Orient Blackswan Private Limited
3-6-752 Himayatnagar, Hyderabad 500 029, Telangana, India
e-mail: info@orientblackswan.com

Contents

Images, Figures, Tables and Appendices

Images

Figures

Tables

Appendices

Abbreviations

ABP	Aravalli Biodiversity Park
ABWS	Asola Bhatti Wildlife Sanctuary
ADC	Autonomous District Council
AFSPA	Armed Forces (Special Powers) Act
AoL	Art of Living
BJP	Bharatiya Janata Party
BNHS	Bombay Natural History Society
CAG	Comptroller and Auditor General
CASI	Centre for Advanced Study of India
CCLA	Chief Commissioner of Land Acquisition
CCMB	Centre for Cellular and Molecular Biology
CEDAR	Centre for Ecology, Development and Research
CEE	Centre for Environment Education
CF	Community Forest
CFM	Community Forest Management
CIC	Central Information Commission
CMFRI	Central Marine Fisheries Research Institute
CPR	Centre for Policy Research
CPR	Common Property Resources
CPWD	Central Public Works Department
CRRF	Central Ridge Reserve Forest
CRZ	Coastal Regulation Zone
CTA	Criminal Tribes Act
CVCA	Critical Vulnerable Coastal Area

CWRA	Central Wetland Regulatory Authority
CZMA	Coastal Zone Management Authority
CZMP	Coastal Zone Management Plan
DBOFT	Design, Build, Operate, Finance and Transfer
DDA	Delhi Development Authority
DoC	Department of Commerce
EAC	Expert Appraisal Committee
EIA	Environmental Impact Assessment
ESA	Ecologically Sensitive Area
FAO	Food and Agriculture Organization
FD	Forest Department
FSI	Forest Survey of India
GCZMA	Gujarat Coastal Zone Management Authority
GIS	Geographic Information System
GoI	Government of India
HTL	High Tide Line
IBA	Important Bird Area
ICMAM	Integrated Coastal and Marine Area Management
IDA	International Development Association
IISc	Indian Institute of Science
IMCLS	Idu Mishmi Cultural and Literary Society
INTACH	Indian National Trust for Art and Cultural Heritage
IUCN	International Union for Conservation of Nature
JFM	Joint Forest Management
JFMC	Joint Forest Management Committee
JHADC	Jaintia Hills Autonomous District Council
KFD	Karnataka Forest Department
KPA	Karnataka Planters' Association
KPTA	Karnataka Preservation of Trees Act
KSIIDC	Karnataka State Industrial and Infrastructure Development Corporation

LPG	Liquid Petroleum Gas
LTL	Low Tide Line
LULC	Land Use and Land Cover
MoEF	Ministry of Environment and Forests
MoEFCC	Ministry of Environment, Forest and Climate Change. MoEF (see above) was renamed as MoEFCC in 2014.
MoES	Ministry of Earth Sciences
MPA	Marine Protected Area
MW	Mega Watt
NATMO	National Atlas and Thematic Mapping Organization
NBWL	National Board for Wildlife
NCC	Nagarjuna Construction Company
NCPC	National Campaign for Protection of the Coast
NCR	National Capital Region
NCSCM	National Centre for Sustainable Coastal Management
NCZMA	National Coastal Zone Management Authority
NDMC	New Delhi Municipal Council
NEC	North-East Council
NEERI	National Environmental Engineering Research Institute
NEF	North-East Frontier Agency
NEIIPP	North-East Industrial Investment Promotion Policy
NGO	Non-Governmental Organisation
NOC	No-Objection Certificate
NR	Northern Ridge
NSDP	Net State Domestic Product
NTFP	Non-Timber Forest Product
NWCP	National Wetland Conservation Programme
PAG	Plantation and Agriculture
PA	Protected Area
PCCF	Principal Chief Conservator of Forests

PDS	Public Distribution System
PIL	Public Interest Litigation
PTI	Press Trust of India
PWD	Public Works Department
RF	Reserved Forest
RS	Remote Sensing
RTI	Right to Information
SANDRP	South Asia Network on Dams, Rivers and People
SCZMA	State Coastal Zone Management Authority
SEZ	Special Economic Zone
SPCB	State Pollution Control Board
ST	Scheduled Tribe
TEEB	The Economics of Ecosystems and Biodiversity
UK	Uttara Kannada
UN	United Nations
UNEP	United Nations Environment Programme
UPASI	United Planters' Association of Southern India
UPIASI	University of Pennsylvania Institute for Advanced Study of India
UT	Union Territory
VP	Van Panchayat
WII	Wildlife Institute of India
WTI	Wildlife Trust of India

Acknowledgements

This book had its origins in informal discussions at the Centre for Advanced Study of India (CASI) at the University of Pennsylvania where Ghazala Shahabuddin spent two months in late 2015 to study the politics of wildlife management in India. Her meetings and discussions with Devesh Kapur, then Director of CASI, led to the idea of an India-based workshop to further explore contemporary issues in wildlife conservation and environmental regulation. Subsequently, after further discussions at Yale University, we decided to take the process forward and crystallised the idea: of studying contemporary issues in nature and biodiversity conservation, particularly the drastic changes in the use, management and regulation of natural resources in India. These issues are especially pertinent when taken up 25 years after ambitious economic reforms were initiated in India in 1991. Accelerated economic growth in the last three decades has created added pressures on natural resources and nature conservation areas. At the same time, new ideas have come to infuse the way projects of sustainable development and species conservation are designed. We decided it would be instructive to take stock of these changes through a consideration of national policies, regional case studies, and the state of scientific research. With these objectives, our workshop entitled 'Wildlife Conservation in India: From Policy to Practice' was held on 4–5 November 2016, at the India Habitat Centre in New Delhi, with support from CASI and the University of Pennsylvania Institute for Advanced Study of India (UPIASI). Yale University kindly provided support for K. Sivaramakrishnan to attend and co-organise the workshop.

We would like to extend our heartfelt thanks to Devesh Kapur for his encouragement and for facilitating the use of CASI's resources for holding the workshop in India. The staff at CASI, especially Juliana DiGiustini, were generous with their time in enabling and supporting the workshop and were enthusiastic about the development of a book from the deliberations. The workshop itself was meticulously organised by the staff at UPIASI at the India Habitat Centre. We would like to specially thank E. Sridharan, S. D. Gossain and Ruchika Ahuja of UPIASI for their thoughtful planning and efficient logistical support.

Papers presented at the workshop covered recent historical developments in conservation policy, as well as contemporary changes in the policy environment. They also examined several cases from the Himalayan region to central India and urban ecology in cities in the north and the southern parts of India, examining conflicts over resources and species endangerment. Meghna Agarwala, Chetan Aggarwal, Ambika Aiyadurai, Rajkamal Goswami, Shashank Kela, Kanchi Kohli, Eszter Kovacs, Kartik Shanker, Rinki Sarkar, Neha Sinha, Abi T. Vanak, Mayank Vikas and Bhaskar Vira presented papers at the workshop. We thank each of the authors for contributing to the engaging and lively discussions during the two days of the workshop. We also thank their co-authors, Manju Menon (with Kanchi Kohli), T. Ganesh (with Rajkamal Goswami) and Chandrima Home (with Abi T. Vanak), who could not attend the sessions but were very much part of the discussions through which this book has emerged. We especially honour the memory of Archana Bali, outstanding student of ecology and conservation biology, who passed away in 2014; she co-authored the paper that Kartik Shanker presented at the workshop. We were also pleased to be able to include the work of a number of authors who had just completed their doctoral studies at the time of the workshop, bringing their new research into focus alongside the contributions of several senior scholars. We were also happy to generate a wide-ranging discussion as papers were presented by social science scholars, as well lawyers, journalists, public intellectuals and environmental scientists.

Rajesh Thadani, Mahesh Rangarajan, M. D. Madhusudan, Arupjyoti Saikia, Shomona Khanna and Nimesh Ved kindly accepted their role as session discussants and reviewers of the papers presented at the workshop. Their in-depth commentaries on each of the papers were critical to improving the quality of the final manuscripts that ultimately became part of this book. Mahesh Rangarajan went out of his way to offer detailed guidance to the overall workshop design and by sending comments in writing to each of the paper authors who participated in the workshop. All these reviews and conversations were important in enabling us to think with this set of papers and come up with a book project that would provide some key ideas and commentary on the state of nature and wildlife conservation in India since the launch of economic reforms and the rapid increase in development activity.

We thank Orient BlackSwan (OBS), and particularly Roopa Sharma, who has been a patient and exemplary editor for us. She has worked with us through many challenges to help us finalise this book. We are grateful to her for the prompt and conscientious handling of the manuscript. An anonymous reviewer read the entire manuscript and helped us refine the final product. The copy-editor at OBS, Pooja Sanyal, has done a remarkable job and her patience and eagle-eye have been instrumental in delivering a high-quality manuscript.

We thank our respective family members for their support through the years that have enabled our intellectual pursuits. I (Ghazala Shahabuddin) would like to thank my husband Jawed Ashraf, and my parents, Rashda Ashraf, late Suleiman Ashraf, Shaher Bano and late Syed Shahabuddin for their unstinting encouragement throughout my career. Through the years, scholars Rauf Ali, Ullas Karanth, John Terborgh, Ramachandra Guha and Mahesh Rangarajan have helped me formulate and crystallise ideas out of a complex morass of diverse and often conflicting, ideas and opinions. I am grateful to each of them for their guidance and help for many years: a scientist who dared to dabble in the social side of conservation, could not have hoped for a more eminent or diverse set of mentors! I (K. Sivaramakrishnan) would like to thank my wife, Saroj Sivaramakrishnan, for her fortitude and consideration as my travel for research and other professional obligations took me away from home a lot over the years, not least for the conduct of the workshop from which this book has now been prepared. Her own love of wildlife, care for animals big and small, and career of tireless service to the ailing poor, is a salutary reminder of the difficult balance that has to be found for all forms of life to flourish on earth.

Ghazala Shahabuddin
K. Sivaramakrishnan

Glossary

aadulo chhin	fermented soyabean
aakru	takin
aala	Himalayan musk deer
aalaapee	musk pods
aambo	a type of generic ritual performed to ask for success and safety
aame	monkeys
aamme	wild boars
aamra	tiger
aanphoodi (also called *paao yute/seekhi yute*)	brass metal
aanphun aangi	a ritual performed by the hunter after a successful hunt
Adiju	the name of a spirit in Mishmi mythology
adivasis	tribal people
ahun	bears
akupi	mushroom
ala	musk deer
ambo	maize
amrala	tiger teeth belt
amrase	a healing ritual usually carried out for a day
amuneenddon	skull boards
anapra	vegetables

angein	strict taboos
anji	hunting
anjiya	hunters
apiya	elder brother
arabica	a type of coffee
awla/amla	fruit of the tree *Phyllanthus emblica*, which is an important component of Ayurvedic medicines
babul	refers to native tree species of genus *Acacia*, found in scrub forests
bahadur	an amicable way of addressing migrant employees
bajra	a type of grain
bander	harbour
bangar	wells
beela	waterbody
behera	tree species *Terminalia bellerica* whose fruit is widely traded for its medicinal value
bhur	sandy loam
bombil	dried and salted Bombay Duck
bund	embankment
chargah	grazing land
chinkara	the local name for Indian gazelles
chital	spotted deer or *Axis axis*
dabar	low-lying catchment area that the Yamuna frequently floods
dakar	clayey loam
dao	machette
devta	deity

dhok/dhau	the tree species *Anogeissus pendula* that commonly grows on hill slopes of the Aravallis
doab	the Indo-Gangetic plains
dolois	a traditional chief
drati	a handy tool for pine cone collection
elaka	usually a cluster of 10–40 villages
elompran	onion
eloni	garlic
fajr azaan	the call to morning prayers
gair mumkin pahar	non-cultivable land
gandanaala	dirty drain
gaon buddas	village heads
gaon sabha	the village community
Gaon Vikas	village development
gazni	salt water rice fields
Golon	spirit that guards wild animals
harra	tree species *Terminalia chebula* whose fruit is traded for its medicinal value
jalau	trees that give good fuelwood
jhand	*Capparis sepiaria*, a thorny shrub growing in scrub forests of the Aravallis
jheel	lake
jhum	shifting cultivation
jogti	pine torches
kan	evergreen forest patches in the Western Ghats in Karnataka
khadar	riverine, sandy land
kheenu	spirits

kohi (also called *pahari*)	ridge (term used specifically for the Aravalli hills)
la-ro	a symbolic gate
lathi	stick
mahua	tree species *Madhuca indica* which is integral to the tribal economy due to its useful flowers and fruits
mandir	temple
marsana	a herb
Mitu-Sipa	spirit that guards domestic animals
munj	*Saccharum munja*; a type of aquatic grass with many uses including as thatch and fodder
Nanyi Inyitaya	the Supreme Creator
nilgai	blue bull, an antelope species
nistaar	forests allotted for meeting daily household needs
pagadia	on foot
pala	a type of bush-weed
pan-jhum	betel leaf
Pnars	the indigenous and dominant tribal community of Jaintia Hills
Prosopis juliflora	exotic (and invasive) tree species also known as mesquite or *vilayati kikar*
rajakaluves	storm water drains
rakh	thorny, secondary forests
rausli	clayey-sandy loam
Reh	a festival of the Mishmis for propitiation of the Supreme Creator
robusta	a type of coffee, derived from the plant *Coffea robusta*

satkatha	a term that refers to the forest as a whole with many species
sckol	cooperative spirit and reciprocal social arrangements
shamlat (also called *shamilat-deh*)	village commons in Delhi
taamaamran	ritual required to be performed if a tiger is killed or trapped
taluk	block
tampara	lotus or lotuses
Tampara Poramboke (or *Beela Gayalu*)	the land of the lotus
tankha	money from land
tendu	tree species *Diospyros melanoxylon* whose leaves are widely traded for making country cigarettes
thekedar	contractors or middle-men
tola	a unit to measure gold and musk pods
waheh chnong	the village headman
yamba	millet
yiu	rice beer

Introduction

Ghazala Shahabuddin and K. Sivaramakrishnan

Conservation and Development

In the new century epitomised by economic reforms and global trade, nature, wilderness, biodiversity and a wide range of resources derived from the natural environment—air, water, fuel and food—are under unprecedented stress in India. This is so, despite nearly five decades of growing efforts to implement laws, policies and programmes for conservation, and to do so in alignment with two other priorities: development and democracy. An examination of this stress and the solutions being found across regions, resources and communities must question the ideas of nature protection that shaped both late-colonial efforts and what followed in the first four decades after India gained independence from British rule. And it must evaluate the ideas of conservation as civic duty, or social enterprise, or ecosystem services, which have been developed over the last 30 years, even as the Indian economy entered a phase of sustained rapid growth. In this sense, this book is a collective assessment of nature conservation after economic liberalisation in the 1990s, for the most part; inevitably some historical processes leading up to this juncture have to be studied as well.

In the late nineteenth century, nature conservation in India had its beginnings largely in the form of hunting regulation by erstwhile princes to preserve game animals or to mitigate conflict with them (Rangarajan 2001). While a few protected areas (PAs) were thus created even before the Second World War, growing importance was afforded to the protection of wildlands after Independence. Project Tiger, initiated in the 1970s, came to exemplify an ambitious national imperative to protect charismatic megafauna, unprecedented at the time in a developing country (Ranjitsinh 2017; Sankhala 1977). However, the rapid expansion of territories sequestered for conservation in India had to await the last quarter of the twentieth century (Ramesh 2017; Rodgers et al. 1988; Wildlife Institute of India [WII] 2017).

Starting in the 1970s, political will and growing middle class support—as well as the rise of environmental movements—spurred improvements in nature conservation measures nationally (Ramesh 2017). The creation of a strong legal and regulatory framework, appropriate to the local, regional and national needs of environmental conservation, was one of the results of this growing political awareness through the 1980s (Sivaramakrishnan 2011). In a recent overview, Mahesh Rangarajan reminds us that the period from 1969–89 was crucial in many respects for the advance of nature conservation in India. He writes, 'the two decades saw not only a five-fold growth in the acreage under national parks and wildlife sanctuaries, but also the enactment of key laws ... ' (Rangarajan 2017). Such an expansion of protection to natural habitats was seen as a major achievement for a developing country and was much ahead of its time (Ramesh 2017).

This central impetus was matched by the engagement of state governments, local bodies and forest-resident communities or those villages that lay in the buffer zones of PAs (Kothari 1998; Sundar et al. 2001). Importantly, decentralised governance through local institutions such as panchayats, municipalities and forest councils was encouraged and these local bodies were empowered through policies enacted in the 1980s and 1990s, giving them much more local control over natural resources than was possible earlier.

While many studies of Joint Forest Management, people and parks programmes, and integrated watershed development schemes have followed, the findings are not very divergent despite disparate methods and often rival theoretical orientations. Yet, engagement with local power and management after 1980 perhaps remained sporadic and largely inadequate. Participatory conservation was limited by the continuing custodial attitudes of government agencies charged with implementing these schemes and the persistent asymmetries of power in village communities that hamper democratic decentralisation and deliberative democracy (Springate-Baginsky and Blaikie 2007). Additionally, in some quarters of the nature conservation world—both in India and among the international sponsors—there remained a deep suspicion of the idea that involving local communities was not compatible with the long-term flourishing of wildlife, especially endangered species. Examples of these sentiments may be seen, for instance, in the dissenting note of Valmik Thapar in the Tiger Task Force report (Ministry of Environment, Forest and Climate Change [MoEF] 2005) or in essays by Ullas Karanth (2006).

While the outcomes of such conservation schemes have remained underwhelming, the experience did mobilise widespread support for more effective participation of local communities, especially in tribal areas, in decisions relating to forest management and use. These struggles culminated in the passage of the Scheduled Tribes and Other Traditional Forest Dwellers (Recognition of Forest Rights) Act, 2006. This law spurred its own controversies, including attacks from several groups who saw it as undermining wildlife conservation. The implementation of this law remains very uneven across different states, even as it has enabled some scattered acts of resistance to the conversion of forest lands for mining projects or other environmentally deleterious uses (see N. C. Saxena 2010; Sivaramakrishnan 2015).

Conservation science also began to play a more direct role in wildlife and forest management in the 1970s; Project Tiger, 1972, has been seen as the turning point. During the decades that followed, scientific research in India was gradually institutionalised in a technocracy. Most of these institutions had their origins in colonial era imperatives of natural resource extraction, notably timber and sport-hunting (Rangarajan 2001). During the twentieth century, state-sponsored historical institutions such as the Forest Research Institute and the Zoological Survey of India were modified to meet the needs of conservation in the midst of competing demands for land and habitats for wildlife. International conservation agencies and the rising field of conservation biology came to influence the work being done in India after the 1950s, although they remained marginal in a largely state-controlled wildlife protection system (Lewis 2004). The WII was created in 1987, in response to the perceived need for a cadre of forest officers who could better understand and conserve wildlife (Ranjitsinh 2017).

Despite huge investments in PAs, conservation science and local institutions, challenges to biodiversity conservation have only grown since the initial successes in the 1980s. India's population has grown multiple-fold, from 568 million in 1971 to 1.2 billion today, with a concomitant expansion of demands for resources. In the face of the needs to reduce poverty and satisfy growing material aspirations, competition for land has increased over the years. Simultaneously, rapidly rising rates of consumption have been engendered by industrial expansion and accelerating urbanisation, and have also demanded widespread infrastructural development since the 1990s. At one level, the pace of economic growth picked up remarkably in the 1990s and seemed to further increase in the following decade. Such prosperity, especially for the middle classes, the wealthy, and to some extent, organised labour, has been expressed as a pressure on land for commercial development, housing, clean water and sanitation, and has led to an exponential increase in waste generation, especially in cities. The intersectoral performance of the economy in India, and thus, the contribution of various sectors also placed added pressure on land, as agriculture lagged behind various service sectors (Balakrishnan 2010: 173–209).

Land speculation has increased; the demand for minerals and metals has grown; energy, water, various fuels, and all manner of communication infrastructure, from mobile phones to transport are rapidly expanding. In fact, such a gamut of economic processes can be likened to the 'Great Acceleration', referring to the period of rapid increase in consumption and economic growth rates globally after the Second World War. On the global scale, it is dated from 1945, and this acceleration is described in terms of the astronomical increase in motor vehicles and the corresponding demand for fossil fuels, urban concentration, and the intensification of demand for primary resources reflected in deforestation, degradation of fisheries, atmospheric pollution, and accelerated mining of minerals, precious stones and rare earths (McNeill and Engelke 2014). In the Indian context, this idea may be more appropriately situated in the patterns of economic growth since the economic liberalisation in 1991. These patterns have also aggravated regional divergences or accentuated uneven development (Sanga and Shaban 2017). Such trends obviously began to exert tremendous pressure upon the remaining natural ecosystems and resources in India, a pressure that is only growing and that has significant negative implications for biodiversity

(Bindra 2017; Shrivastava and Kothari 2012). Since PAs often fall in relatively disadvantaged economic regions, in the Himalaya or the far northeast, or central India, growing wealth disparity within regions of the country increase degradation stress in these PAs. Wealth disparity increases since the 1990s have also concentrated and magnified the economic and political power of some of the richest and socially dominant groups in Indian society (Anand and Thampi 2016).

Numerous problems associated with accelerated development —land use change, forest loss and degradation, solid waste proliferation, and water contamination and scarcity—today threaten to overwhelm the gains made in protecting biodiversity since the 1970s (Bindra 2017; Shahabuddin 2010). Available statistics confirm growing crises for protected species and habitats across different ecological zones in India. Each year several species are added to the threatened list; forest cover declines and quality of habitats and ecosystem services degrade further—freshwater, marine and terrestrial. For instance, the population of the critically endangered Great Indian Bustard, an endemic bird species, which lives in semi-arid grasslands, has been plummeting, despite several dedicated wildlife reserves (Bindra 2017; Rahmani 2003). It is one of the several species in India predicted to go extinct over the next few years. Loss of riverine habitat and pollution afflicts the Ganges gharial (Lenin 2007); the olive ridley turtle suffers on account of development along its nesting beaches in eastern India (Bindra 2017); and the common birdwing butterfly has declined due to loss of dense hardwood forests in the Himalayas (Smetacek 2012). Such relatively unknown species may disappear as their problems have been ignored in the cacophony of voices over the conservation of more attractive species such as the Asian elephant. Therefore, even with some success in conserving charismatic megafauna like the lion or the tiger, and considerable attention drawn towards the plight of elephants and rhinos, many other species of animals, birds, marine life and invertebrates are declining as habitats are disturbed and destroyed. Large herbivores such as elephants are threatened much more today by linear infrastructure that disrupts corridors and fragments habitats than the presence of small forest settlements or marginal cultivation (MoEF 2010). Arguably, the losses of biodiversity are most acute in areas targetted for large infrastructure projects like ports in coastal areas, mining in forest areas, and industrial and large housing developments in wetlands, marshes, and along fresh water bodies like rivers, lakes and tanks.

Along with biodiversity, many of the services that are delivered by robust and well-functioning ecosystems are increasingly threatened (The Economics of Ecosystems and Biodiversity [TEEB] 2010). While the net area of forest in India has remained more or less constant or even increased patchily, there is visible decline in the quality and diversity of native forests (Shahabuddin 2010; Singh 2015). Ecosystem degradation directly affects quality and regeneration of plant species, especially trees; the economic losses due to such decline have been estimated by economists to be in the range of several trillion dollars (TEEB 2010). Availability of forest products to people, clean water flows, disaster mitigation, hydrological services and carbon sequestration have been adversely affected due to biodiversity loss, with concomitant effects on local and regional economies. Arguably, the quality of life in polluted and resource-stressed cities, as well as commons-deprived villages, has declined in the last few decades, not least due to degradation of forests and wetlands. Even as the economy has grown, social inequality has been exacerbated, making resource struggles more acute and rancorous, and often, leading to stand-offs between primarily rural, ecosystem-dependent people and urban votaries of industrial and infrastructural growth (Shrivastava and Kothari 2012). The visibly escalating conflict between conservation laws and the need to provide natural resources for development seems to have limited pathways for resolution today.

The proliferation of regulations, laws and projects for conservation from 1980 till today, while unique for a developing economy, did not always result in the spread of systematic environmental management. The Wildlife Protection Act, 1972, the Forest Conservation Act, 1980, and the Environment Protection Act, 1986, are detailed and comprehensive, but their implementation has not only been highly inadequate, but also, often selective, sometimes exhibiting a bias towards more influential players such as large corporations and the urban elite. Further, the range of laws has not created the broad public sympathy for wildlife that they were expected to generate. Instead, the frequent subversion of people's rights locally, even those mandated by the Wildlife Protection Act, left many of the wilderness areas and bordering landscapes without protection of any kind. Escalating human–

wildlife conflicts, which are caused by crop-raiding or livestock-killing, result in substantial losses to agriculturists and pastoralists and further reduce public support for conservation (Madhusudan and Mishra 2003; Shahabuddin et al. 2007). In this scenario, a lot of the laws and policies that have been put into place for ecosystem protection have been rendered ineffective; often they are outright unpopular, and thereby, lacking legitimacy.

The large-scale failures of conservation laws and policy prompt an inquiry into why they do not work effectively on the ground. Are there failures at the policy-making stage that reflect an inability to take into account ecological and socio-economic realities, or are there lacunae simply in implementation of what might well be fine policies? How have people, governments and scientists been addressing biodiversity challenges, individually and collectively, within and outside the legal framework and how successful have they been? Is the state of scientific knowledge sufficiently advanced and adequately included in decision-making to play a more central role in policy formulation than it does currently? What are the historical roots of some of the ecological conflicts and also governmental responses that constrain wildlife conservation in current times?

This volume, building on sustained case research from different parts of India, presented at our workshop in 2016, endeavours to understand the institutional, legal, scientific, political and social constraints in meeting conservation goals in India. The fast-changing socio-economic canvas and new ecological stresses created by economic liberalisation in India since the 1990s form the context to view and study these challenges.

Specifically, this book is unified by an interest in the way economic liberalisation created added pressure on PAs and species and led to new approaches to conservation, where protection by both exclusion or community-based conservation, have been re-imagined as the conservation of valuable ecosystem services, often by social and private enterprise. In this sense, we are seeing a commercialisation of the goals and methods of nature conservation and a monetisation of its likely benefits and costs. This examination has required learning not only from both science and social sciences, but also fields like ethics and the humanities to assess how conservation becomes a project for corporate enterprise or civic responsibility, as a broader environmental ethic or inter-species understanding. This also necessitates an investigation of how the natural and social world permeate common perceptions among different classes of society in India in the wake of new prosperity, for some, and grinding poverty and greater vulnerability, for others.

A Multi-Disciplinary Perspective

Given the existing literature, we believe that any significant understanding of a complex field such as nature conservation—which cuts across the boundaries of the biological and social realms—cannot be dealt with adequately within the purview of a single discipline. A deeper treatment of the subject of nature conservation necessarily involves dialogues across several areas including ecology, anthropology, history, political science and even economics. This is especially true when one attempts to analyse the real life consequences of such policies, comprehend the cultural processes in which environmental values are formed and change, or seek solutions to ecological problems within the socio-political realm in which all conservation must ultimately be validated and implemented.

Two broad approaches and associated styles of writing can be discerned today in the subject area of nature and wildlife conservation. Much of this work is firmly divided along disciplinary lines. The first genre, which is most prominent in public debates, focuses on the biological issues in conservation, and has been populated mainly by naturalists and natural scientists. Most often such work has emanated from a single species perspective—elephant, tiger, bear, turtle or antelope (for instance, see Karanth 2006; Thapar 2001); but occasionally it has also generated work on distinctive and rare habitats like rainforests or arid zone ecology, as well as much smaller and diverse groups (for instance, see Dharmakumarsinhji 1998; Krishnan 2012). In a parallel stream of work, mostly carried out by social scientists, the research springs mainly from concerns of economic disparity and livelihood disruptions, environmental politics, social movements and human ecology. This list includes authors such as Madhav Gadgil, Ramachandra Guha, K. Sivaramakrishnan, Mahesh Rangarajan and Arupjyoti Saikia (for key publications as such, see Rangarajan and Sivaramakrishnan 2013).

There are still very few books that expressly attempt to analyse and understand nature and wildlife conservation in a way that adequately bridges the biological sciences–social sciences/humanities divide. A few volumes in recent times come to mind: *Battles over Nature* (Saberwal and Rangarajan 2003), *Nature Without Borders* (Rangarajan, Madhusudan and Shahabuddin 2014), *People, Parks and Wildlife* (Saberwal, Rangarajan and Kothari 2001) and *Democratizing Nature* (Chhatre and Saberwal 2006). Most recently, *From Soup to Superstar* (Shanker 2016) and *Conservation at the Crossroads* (Shahabuddin 2010), unusually for single author volumes, do manage to connect the science and politics/policy of conservation effectively. Monographs such as those by Michael Lewis (2005) have illuminated extremely well the machinations by which even individual politicians and bureaucrats can decide the fate of wildlife, both ecosystems as well as individual species. Prerna Bindra's recent book, *The Vanishing* (2017), has focused on the question of development versus wildlife, showing how the needs of biodiversity are being continually sidelined in the course of economic development, a process which has been dominated by accelerated infrastructural development and industrial growth in India during the last three decades.

From a historical perspective, too, this volume is expected to occupy a special place. There is very little published material that explicitly deals with the period of economic liberalisation from 1991 and goes into contemporary situations. During the last three decades, significant political, social and economic changes were unleashed throughout the country via economic reforms. The debates and institutional arrangements for conservation have also become more diversified in this period. Arguably, the courts have taken a more prominent role, government agencies have been redesigned and expanded, and have worked more closely with civil society, where conservation-focused non-governmental organisations (NGOs) have grown in numbers and financial resources. Most important, by their greater involvement and often distinct perspectives on nature and its identity related to provinces or states, regional governments and regional elites have made conservation no longer a preserve of metropolitan elites and national policy-makers (see, for instance, several essays in Cederlof and Sivaramakrishnan 2006).

In this volume, we also note the expansion of conservation laws in the last few decades of the twentieth century, precisely in the historical period when such centralisation of government was being challenged by political processes that had generated growing demands for local and regional self-government. A shift in the way citizenship is understood and experienced is also noticeable in the spread of rights legislation post-1990s, which seeks to entrench social and economic claims on the State in a slate of laws such as the right to information, food, forests, or education. This volume takes the spread of rights-based approaches to poverty reduction and social inclusion in India seriously even as it examines the growth of conservation bureaucracies and programmes that do not seem to be well-aligned with the rights revolution.

In doing so, we have to begin with a close examination of forest governance as it developed in the twentieth century. At the best of times, biodiversity governance showed scant regard for both conservation science and the demands of environmental activists or social justice movements fighting for *adivasis* in and around forest areas. The early decades of conservation focused on PAs, and even when more inclusive approaches were adopted, they rarely produced any necessary reform in the working of the forest department, which however, remained the custodian of all forest lands where wildlife, biodiversity and forest conservation was enacted.

Wildlife Laws and Policies

As mentioned earlier, 1980–2015 represents the period, starting with the passage of the Forest Conservation Act, 1980, and then the Environment Protection Act, 1986, which witnessed an expansion of conservation activities and associated policies and regulations. And, to further deepen the importance of these 30 years, the constitutional amendments mandating regularly elected local governments in the form of Panchayati Raj in rural areas and municipal administration in cities and towns also emerged in the early 1990s, giving previously unknown power and stability to the elected local government. In the midst of these political developments, conservation programmes grappled with both participatory and more exclusionary approaches to a growing network of PAs for charismatic megafauna, endangered plants, animals and habitats (Rangarajan 2015; Thapar 2001). Thus, the period since the economic reforms of the 1990s is interesting to observe in terms of the interplay of federal politics, rapid socio-economic change and highly stressed habitats, endangered species, and vulnerable ecosystems.

In the larger picture, the federal structure of the Indian government necessitated the associated division of legislative responsibility between the central government and states on subjects like land management, rural development, infrastructure, urban planning and the rule of law. From the time of Project Tiger (1972–73), these divisions created an undercurrent of tension between the economic priorities of state governments and the national goals of wildlife protection, which were more often imposed by a strong centre. Wildlife conservation, being a concurrent subject, was increasingly seen as a project designed and driven by centrally enforced elite politics and a vanguardist approach taken by scientists and environmentalists (see, for instance, Guha 2003). This approach was enabled and embodied in a strong Prime Minister, Indira Gandhi, for many years (Ramesh 2017). Yet, the resistance by state governments to the central wildlife laws continued and was later visible in an increasing number of denotifications of PAs in the 2000s, and in the frequent flouting of environmental norms in infrastructure and industry projects. Setting up of new centralised authorities such as the National Tiger Conservation Authority, created in 2005 after the local extinction of tigers in Sariska, Rajasthan (MoEF 2005), seemed to suggest that state governments were thought less and less capable of safeguarding species in the face of economic imperatives. These contradictions are present to this day and are no closer to getting resolved.

Specifically, this volume offers some insights into how both democracy and federalism shape the legal and rights-based approaches to natural resources management and conservation in a shifting landscape where the Forest Rights Act, 2006, is now a decade old and where significant revisions are being considered to a nearly 50-year old Wildlife Protection Act, 1972. Our case studies also examine the implementation of specific laws and rules, as they respond to international expectations, national priorities as well as local constraints.

The Wildlife Protection Act, 1972, for instance, has been the central, overarching law meant to safeguard ecosystems from complete modification or conversion for human use, through the establishment of no-go areas (PAs) and to protect native species of flora and fauna from harm. Yet, this very law that protects species and habitats has little or no traction in some of the north-eastern Indian states, where hunting is widely prevalent, being inextricably linked to the subsistence and culture of local tribes. In fact, the north-eastern region of India is often seen as a lost cause for conservation by most Indian conservationists, as hunting (for subsistence and sport) by tribal people is thought to significantly deplete wildlife. Several highly endangered species have been caught in this crisis as well.

In Chapter 1, Ambika Aiyadurai explores the complex social and cultural factors for the lack of acceptance of the Wildlife Protection Act among the Mishmi people in Arunachal Pradesh, which leads to its poor implementation. She shows how the average conservationist has failed to understand the complexity of the problem, and with a certain degree of insensitivity, has supported the blanket ban on hunting mandated by the Wildlife Protection Act. She argues that in fact, the very centralised and black-and-white nature of the conservationists' perception of hunting may be an obstacle to effective wildlife conservation in this region. Aiyadurai urges the reader to engage with the cultural and economic aspects of hunting in a deeper way, in order to come up with alternative means of dealing with the issue, which negotiate carefully between people's needs and those of biodiversity conservation. Some options that can be explored, for instance, include involvement of the local government in PAs through the legal category of Community Reserves, nature-based tourism or compensatory payments for protection. Sustainable offtake of herbivores, such as the wild boar or nilgai, whose populations have reached damaging proportions in some areas, is also an oft-discussed possibility. But such an idea has met with many critics in the conservation world; the lack of a biological basis for shaping the nuanced policies precludes useful discussion in a situation where contrasting positions are often sharply opposed.

Another facet of the Wildlife Protection Act is the uniformity in its application to the establishment and management of PAs in India. This lack of a nuanced and variegated implementation of the law, which points to both rigidities in the statute and the lack of discretionary power in the hands of state agencies, allows little room for adjustments based on local requirements—ecological or social (see also MoEF 2005). Ecology and social structures obviously vary dramatically from one region to another, apart from the socio-political/historical contexts in which conservation is undertaken. In Chapter 2, Mayank Vikas shows how the pastoral village communities that lived off the scrub-savannahs of Delhi were criminalised overnight by a notification of the Asola Bhatti Wildlife Sanctuary in south-eastern Delhi. While people were allowed little leeway in the use of resources from the PA, several state-sponsored activities were gradually allowed and even legalised, such as trash dumps, housing projects, metro-lines and recreational parks, all of which do more irreversible damage than grazing or lopping. The author illustrates the irony of the implementation of bans on residence and grazing, while on the other hand, widespread changes and impoverishment of Delhi's forest ecosystems took place under the watch of planning experts and bureaucrats (see also Agarwal 2014 on Delhi Ridge forests). Such inconsistencies also produced a sustained and widely joined protest against the degradation of the Delhi Ridge and a series of court cases that led to the reimagining of the Ridge forests as the source of varied ecosystem services in Delhi. Moving from nature conservation for species or landscape protection to the concept of ecosystem services crucial to urban environments has apparently broadened the debate in this case (Sivaramakrishnan 2017).

Vikas strongly emphasises the need for first acknowledging the various forms of modification and destruction that are going on in Delhi's forests, and then integrating local voices and requirements into the PA system, something that has been long overdue (Lewis 2003; Shahabuddin 2010). This chapter thus points out the inadequacy of the laws that govern, as well as their misinterpretation and selective implementation, privileging certain social groups over others. Such a process is seen in more than one domain, whether it be coastal regions (Kohli and Menon, Chapter 6, this volume), where ports take precedence over fishing interests, or urban wetlands, where recreational needs trump subsistence activities such as grazing and fishing (Nagendra 2016).

Chapter 2 also illustrates the stark shortcomings in conservation policy with regard to protection of the larger rural landscape adjoining or connecting PAs, which is made up of agricultural fields, grazing lands and intensively used forests. Such land is governed under a range of management regimes from privately owned to village commons and state-owned public lands. Such areas, often undefined in terms of exact ownership, provide connectivity, foraging resources and spillover areas for endangered wildlife, and along with the strictly preserved sites, form mosaics of habitats that support and even enhance the survival of wild species (Athreya et al. 2013; Osuri et al. 2010). It is this 'grey' area that is now under deep threat from suburban sprawl and industrial expansion.

Recently, provisions have been created in the Wildlife Protection Act to ensure the continuance of such productive mosaics in the larger multiple-use landscape, such as through allowing for Community Reserves and Conservation Reserves, which can be at least partially, multiple-use areas. However, since this amendment of 2003, only 26 Community Reserves have been established; even these are not very useful, due to lack of outlining of the institutional processes and rules that will be necessary to operationalise them (Shahabuddin 2010). To some extent, the provision for Ecologically Sensitive Area (ESA) (Kapoor et al. 2009), which can be leveraged to create buffers around PAs, also allows for such arrangements, and ESAs have been established in a few places (ibid.). Yet, the enormous difficulty of regulating commercial activities, the powerful lobbies to be countered, the levels of investment needed in human resources, and technical expertise needed to enforce such rules even inside ESAs have led to a situation where the authorities would rather let the 'sleeping dogs lie'! As a consequence, ecosystem-dependent people are squeezed between the loss of erstwhile common pool resources to development, and restrictions in PAs, only leading to further alienation from nature and wildlife.

Some of these issues are brought out vividly by recent research, which is holistic in its scope. For instance, in Chapter 4, Agarwala, DeFries, Jhala and Qureshi have studied forest use by villagers living in the vicinity of wildlife; specifically they have explored the role of fire control and long-term extraction in local livelihoods and the ecological character of forests in the Kanha-Pench conservation corridor in central India. They show how both intended and unintended effects of human use and interaction in these forests can drive changes in quality and species distribution. For instance, resistance to fire and browsing are principal factors determining the survival of saplings into small trees. Interestingly the valuation of individual species by villagers also had an effect. Thus, those valued least were most likely to decline, questioning the easy assumptions that extensive use of certain trees causes their disappearance. Seemingly the complex interplay among use of ecosystem services, rural livelihoods and forest character not only offers no easy answers for conservation, but also allows for spaces that can be exploited for long-term sustainability of nature conservation measures.

Taking a broader and inter-scalar approach to conservation, as the previous example suggests, also invites attention to the interaction between policies that may or may not explicitly be targetted to specific conservation outcomes. Sometimes, policies that seemingly have no connection with wildlife or natural ecosystems have unexpected effects on biodiversity. For instance, Home and Vanak, in a paper presented at our workshop,[1] have found that stray dog populations are expanding in response to the protection they are afforded by the Prevention of Cruelty to Animals Act, 1960. Today dogs are spilling over into adjoining forests and wetlands, where they hunt down vulnerable wildlife, including small antelope, deer, vultures and cranes, and compete with wild carnivores such as foxes (see also Home et al. 2017). However, the Act prohibits any intervention to curb stray dog populations (other than sterilisation), even if they attack and prey on endangered taxa such as snow leopards and great Indian bustards. Much more research is required on such laws, unrelated to environment, which can have significant implications for wildlife.

Apart from overarching central laws such as the Wildlife Protection Act or the Prevention of Cruelty to Animals Act, state-level conservation laws can also be less than effective. Such complications, reportedly common to other Indian states as well, are elucidated in Chapter 9 by Bali and Shanker, where they examine the implementation of the Karnataka Preservation of Trees Act (KPTA, 1976) in the coffee plantation–rainforest mosaic of Karnataka's Western Ghats, an area rich in wildlife due to synergistic effects between these two land uses. The KPTA regulations, however, forbid the felling of native trees even on private land. As a result of this law, plantation owners plant more exotic timber species such as silver oak, as shade trees for coffee, a species they can use and plant without bureaucratic interference. The authors succinctly show that a law that intended to protect native species is actually having the opposite effect: their large-scale decline in the larger landscape. Thus, conservation policies can have perverse consequences for wildlife, primarily due to a lack of understanding of local economies, practices and interactions with wildlife.

Regulatory Environment

Three decades of rapid economic growth since 1990 have seriously enlarged the demand for mineral, land and water, and escalated energy shortages (Shrivastava and Kothari 2012). The quest for these resources have brought development and wildlife conservation into a serious confrontation, as mines, highways, power plants and industrial corridors encroach, pollute and endanger national parks, reserved forests and conserved water bodies (Bindra 2017). Such development, more often than not, impinges upon ecosystem function in a way that influences many aspects of agriculture, forestry and even urban ecology. What is contested is not the act of development per se, but the haphazard and unplanned way in which it is often carried out, not allowing for minimisation of the impacts or discussion of alternative sites. For instance, the Environmental Impact Assessment processes created after the passage of the Environment Protection Act, 1986, are now being diluted to hasten clearances for infrastructure projects, and thus, undermining the environmental democracy that was slowly being built in the preceding decades (see Menon and Kohli 2015). Such acceleration in India arguably coincides with the spread of conservation programmes, bringing resource consumption and conservation into direct opposition in the last three decades.

While a regulatory framework has been put in place to avoid such contradictions, it is clear that the policy imperatives are tilting in favour of the corporate sector and needs of the urban elite, particularly in a situation where it is difficult to ensure independent assessments (Bindra 2017; Gadgil 2001). One of the aims of this volume is to assess these confrontations, which are growing more heated by the day, and the challenges they pose to both conservation policy and practice. In particular, we are interested in the shifting emphasis from law and administrative policy to legal disputes and court-mandated actions as a medium and source of impetus for balancing the needs of sustainable development (see Sivaramakrishnan 2011; 2015). In fact, in a series of case studies dealing with wetlands, coastal zones and other water-related environments, as also in urban environmental conflicts, we are observing a growing role for higher courts and the newly created National Green Tribunal. Following the work of the Green Benches in the Supreme Court of India, the creation of special courts for environmental matters expresses a judiciary-led elevation of these issues and the greater role courts now play in environmental disputes in India (Amirante 2012).

The Wetlands Policy is a case in point. For a very long time, wetlands were not recognised as ecosystems to be dealt with in their own right or to be managed in any concerted way, as elucidated in Chapter 5 by Neha Sinha. Their important role in flood control, groundwater recharge, biodiversity and livelihoods, in addition to many other aspects of national and local economy, has largely gone unacknowledged. This may be attributed to the fact that most wetlands generally exist out of sight of the average urban citizen or, when observed, are primarily perceived as obstructive, disease-harbouring sites (see also Nagendra 2016 on wetlands in Bengaluru). Further, the majority of direct users of lakes, riversides, estuaries and marshes in the city are fisher or artisanal communities, who tend to be marginal to political processes that might determine environmental policies. Other underprivileged groups, sheltered in slums, often occupy the banks of urban rivers and wetlands, testifying to the overall neglect of such systems.

Yet, wetlands are important for the biodiversity they harbour as well as their conservation needs. The land use in a considerable area around (the catchment) is actually more important to their continued existence than purely *in situ* processes, and demand much more scientific and managerial attention than we are willing to spare. Sinha explores the policy surrounding the recognition and management of wetlands and their catchments in India. She shows that the ecological processes maintaining wetlands and their functions, and those that degrade them have not been understood by citizens, policy-makers or even the judiciary, so that an overwhelming majority of wetlands do not belong in the legally encompassed domain of regulated land uses and conservation programmes, with any degree of protection.

Thus, wetlands are gradually being swallowed up by suburban sprawl, solid waste disposal or are wilfully reclaimed (Nagendra 2016). Worse, when wetlands do receive attention, the exploitable commercial aspects often get priority over the need for maintaining the ecological processes. Thus, for example, recreational boating takes precedence over fishing, ignoring the fact that water quality and faunal diversity will be much better served by low-key subsistence uses rather than tourist use, fencing or concrete containment. A major recent development is the role increasingly played by the judiciary in determining the regulatory framework for conservation in India. In the case of wetlands, it has assumed an important role. As a result, the interpretation of various laws by the judiciary really decides which way the pendulum will swing. Thus, the need for a technical understanding of wetland structure and function and the integral, even vital, connection between wetland and catchment health will likely play a more important role in decision-making in future.

Another growing area of conflict as far as environmental regulation is concerned is the rather understudied coastal zone. The coastal zone is the place where 14 per cent of Indian citizens reside and make their living and survive despite the ravages of rising sea levels and storm surges, saltwater intrusion and increasing incidence of coastal storms. The importance of this ecosystem to living organisms as well as ecosystem functions cannot be overstated either, with mangroves, coral reefs, tidal waters, beaches and sand dunes coming under its purview. In Chapter 6, Kanchi Kohli and Manju Menon trace the history of coastal regulation, using the establishment and running of the Coastal Zone Management Authority (CZMA), set up as one of the cogs in the wheel of regulation. They show that the CZMA, set up for regulation of residence and alteration in the coastal zone has become yet another ineffective authority, with a record of repeated dilution of rules over the years.

The history of the CZMA raises the question of whether the current regulatory structures set up for the conservation of the coastline and its associated ecosystems can work at all, given the lack of processes to ensure participation of local communities or to include scientific considerations. As in many other cases, courts have stepped in, where mechanisms developed to facilitate community consultation and the incorporation of appropriate scientific knowledge in environmental decision-making have been ignored or distorted. The example of public hearings in environmental clearances is salient here, as they remain a court-endorsed mode of conflict management in balancing livelihoods, habitats and the growing demands of fast-paced economic development (Mohan and Pabreja 2016). The erosion of the deliberative and learning aspects of such public hearings renders them more of a ritual performance intended to mechanically complete a prescribed process of regulation rather than to integrate science and local perspectives for the best possible sustainable solutions.

Institutions and Social Change

Of particular interest to biodiversity is the role of institutions—both state-level as well as highly local—which have shown inconsistent trajectories lately. A trend that is increasingly becoming prominent is the expanding role of NGOs, and civil society and grassroots activist groups, which attempt to take on the role of conservation research and implementation, thus filling the vacuum created by the lack of dynamism and responsiveness within state agencies tasked with natural resources management. Examining the role of civil and state institutions in a rapidly changing fabric of ownership and market access is, therefore, an important part of our project.

Community-based conservation has recently had a fillip due to increasing recognition of the effectiveness of common property resources and their management. Globally, the *pros* and *cons* of community-based conservation have been much debated: the effectiveness of such institutional arrangements really depends on the specific indicators that are used, as well as the temporal and spatial scales that are considered (Porter-Bolland et al. 2012; Rao et al. 2016). A vast literature drawing on three decades of projects in India as well as sub-Saharan Africa and Latin America has found that community engagement in these projects is limited, often side-lined by local elites, and only occasionally well supported by the government agencies charged with their implementation or oversight. Even in robust democracies like India, where the 1990s saw a renewed commitment to elected local self-government, such participatory conservation and sustainable development projects have suffered due to poor inclusion of all social segments and weak political motivation to enhance democratic government beyond some limited functional decentralisation of agencies.

In India, despite these limitations, the rise and reach of community institutions since the 1980s, however, has been spectacular, as witnessed by the prevalence of Joint Forest Management (JFM),

Community Forest Management (CFM) and Van Panchayat (VP) systems (Shahabuddin 2010). However, in a rapidly urbanising country, and after several years of agrarian distress, rural and forest-dependent people find their long-standing attachments and traditional knowledge related to forest and field being disrupted. This is manifested in the weakening of community-formed institutions at the local level, visible even in relatively less stratified tribal or *adivasi* communities (ibid.). These, after all, are the social and ecological landscapes where mutualism within society and across social and natural systems was higher and more critical to collective survival. Given the relentless focus on these landscapes, however, as stores of natural resources for regional or national development, institutions of mutual preservation that operate more locally can be fatally undermined.

When the rooted understanding of ecosystems is weakened, the institutions that are supported by natural resources will also become unstable and less able to renew themselves over time. In Chapter 7, Rinki Sarkar explores this socio-economic change using the village committees of Himachal Pradesh as an example. Village committees have always been proxy forest managers for the sustainable extraction of pine nuts from the chilgoza pine, a valuable product in high-altitude Himachal Pradesh, and one of the very few sources of cash income. Sarkar documents a rapidly increasing demand for this high value product, soon after roads came to this region, due to improved market linkages and increased consumer demand. Issues of sustainability have consequently come to the fore, seen in compromised regeneration of the endemic chilgoza pine, increased damage to trees and soil erosion. The weakening of local institutions, which were heretofore the primary controllers of the resource, seems to be an outcome of improved market linkages and enhanced connection with mainstream aspirations.

In Chapter 8, Goswami and Ganesh touch upon a landscape much further from the political consciousness of the average Indian, located in a remote part of a global biodiversity hotspot. They explore the forest loss and degradation in the Jaintia Hills of Meghalaya between 1991 and 2011 in both community-managed and state-protected rainforests. Meghalaya is one of the 'seven sisters' state, which remained largely untouched by the developmental economy during most of the twentieth century apart from a low level of artisanal mining. Post 1990s, the growing need for limestone (a necessity for construction and the real estate business) catapulted parts of Meghalaya into a situation of rapid forest loss in the twenty-first century. The authors find that both state-protected reserved forests and community-managed systems suffered forest loss during the period 1990–2010; however, community forests suffered far greater losses than state forests in terms of the proportional area lost. Further, the types of land use change under the two management regimes differed to a large extent. State forests have been largely protected from industrial and mining development, although still experiencing small-scale conversion to monocultural plantations and traditional *jhum* sites. However, the loss of community-managed forests was primarily due to denotification for limestone mining and setting up of related industrial infrastructure. Thus, economically motivated extraction has led to large-scale forest loss and degradation in an area now under pressure for minerals and cash crops, even in some areas controlled by indigenous communities.

Thus, this volume showcases two examples of how, despite traditional controls and autonomy, the working of even remote community institutions has been disrupted by the growing power, greater connectivity and proximity of market forces (see also Rao et al. 2016; Shahabuddin 2010). There are obvious implications for resource sustainability in common property resources, and survival of traditional ecological knowledge, both of which are intertwined with the preservation of the social capital of these village communities. The success of Himachal Pradesh in promoting economic development strategies both relating to agricultural and horticultural development and associated improvements in infrastructure like roads and market hubs might have hastened the decline of community institutions for chilgoza pine forests, triggered, in turn, by the changing labour scenario. Migrant workers take on the work of local communities as their members look elsewhere for work and better prospects. To some, Himachal Pradesh is a success story in inclusive development (see Das et al. 2015). However, it is not clear if even such socially inclusive development in Himachal Pradesh has been beneficial for nature conservation.

At the other end of the pyramid, of no less importance is the role of state institutions of forest management, a topic elucidated by Shashank Kela in our workshop.[2] Barring the early post-Independence years, there has been a visible decline in the quality and credibility of both forest research being done in Indian conditions and the training and motivation of forestry services to best deliver nature conservation in alignment with social justice and sustainability objectives. It may well be that years of inadequate training and expanding expectations have reduced the efficacy of the state forest departments in scientific management, community outreach and even ecosystem protection. Shashank Kela shows that ignoring the needs of specialised knowledge in wildlife and forest management, as well as the lack of the institutional structure required to appreciate and assimilate this knowledge, has led to a visible crisis in forest governance.

Role of Science in Conservation

In the context of governance, the ways in which developments in the conservation sciences are entering policy formulation are of interest, as much as the ways in which spreading knowledge of conservation science in civil society and the media spark new directions for public debate, social action and environmental activism (see also Chhatre and Saberwal 2006; Rangarajan 2015).

Conservation science has come a long way from the period when 'the authoritarian biologist' (Guha 2003) largely had a negative view of local people's role in conservation and focused on creating and maintaining people-free spaces for wildlife. The old-school biologists, many of whom were well-connected to political leaders, and had an inborn enthusiasm for wildlife, tended to fall in with the diktats of the forest departments for strict nature protection. For instance, fire suppression was strictly adhered to, indiscriminately. However, the years after 2000 have seen an explosion of wildlife research that is far more sensitive to local issues of historical use, rights regimes, disenfranchisement issues and to the costs of wildlife protection for local residents; neither are local residents romanticised in their role as the only protectors or guardians of nature, with many studies elucidating anthropogenic impacts on wildlife. In terms of the larger narrative of wildlife decline, large-scale development and local uses are treated equally as destructors; there are no exceptions. Thus, today, most research exhibits a much more nuanced view of people and conservation. Seemingly, the research community in India may be way ahead of its counterparts in the developing world, in its sensitive handling of complex people–nature interactions.

A few realisations related to Indian forests have led to this maturation of the ecological sciences in India. One is that many of the ecosystems previously considered to be pristine have had significant imprint of humans, either in the remote past, or even more recently (see also Hecht et al. 2014). Thus, the green stretches of forest in the Aravallis of Delhi are largely results of planting by the British in the post-1857 era (Agarwal 2014), and the thorn scrub savannahs of central India can be partially attributed to their maintenance by nomadic graziers using fire as a management tool. Second, one cannot justifiably ignore the needs of people living in and around wild habitats, particularly when such people tend to belong to underprivileged and disenfranchised social groups. Third, it is evident that the best research questions can arise and be explored properly only when an ecosystem is viewed and understood holistically—along with the humans within them—given the complex interplay of species and ecological functions in a modified land use mosaic. For instance, with improved research tools at their command, ecologists are finding significant movement of carnivores, such as tigers and leopards, in and around human settlements, a spatial pattern suggesting as much coexistence and cohabitation as confrontation (Athreya et al. 2013; Carter et al. 2012). Finally, natural ecosystems, too, go through much cyclical or linear change naturally, without much human intervention as in the case of cycles of forest pest infestation and the effect on tree populations, for instance. Such changes cannot be easily attributed to human interventions, especially if a long-term view is taken (also see Saberwal et al. 2001). With a burgeoning of research in the areas of ecology and conservation, which take such issues into account, many sites are consequently now much better understood in terms of human influences, evolutionary ecology, and natural patch or disturbance dynamics, creating possibilities for much more effective management in many ways.

But what are the avenues by which such new (and often revolutionary) understanding of many ecosystems can be incorporated into decision-making in India? Historically the forest/wildlife sector has been averse to outside forms of knowledge and has discouraged knowledge-making outside of the state sphere (Lewis 2005). Even today, independent and university researchers, non-profits and think-tanks are thought to be detrimental to the protection of forests and wetlands to the extent that research permits are delayed, withheld and even refused (Shahabuddin 2007). Research that requires interventions of habitat manipulation, experimental instrumentation or animal-handling is next to impossible for the average researcher, as these activities are seen as destructive by the forest departments. Access to remote sites, particularly in the border territories is highly controlled and old maps remain inaccessible. Foreign collaborations are difficult and cumbersome, with many questions of national security being brought up on every application by a scientist to involve an institution or relevant experts, even from leading centres of research abroad (ibid.).

Further, when ecological data is made available by scientists, it is more often than not disregarded in decision-making on the management of ecosystems. Thus, many of the issues surrounding wetland conservation are not acknowledged in the Wetland Policy; much of the data available on coastal ecosystems are ignored during Environmental Impact Assessments (EIAs), and PA managers refuse to take cognisance of and manage well-documented anthropogenic pressures inside their sites. A similar scenario emerges from Chapter 3 by Ghazala Shahabuddin, who analyses the science and politics of the controversial reintroduction programme for the Asiatic cheetah in the Kuno-Palpur Wildlife Sanctuary in India, formally proposed by the Indian government in 2011. While this plan has been judicially stayed (2016) for the time being, owing to court cases, the intriguing series of events that led up to this state-sponsored project illustrates well the tensions among the constituencies of science, politics and local communities in a typical protection scenario in India. This chapter suggests that the planning process overlooked many of the ecological and socio-economic concerns emanating from the possible reintroduction of a large carnivore in a human-dominated landscape from where it had been missing for over half-a-century, leading to a half-baked plan with a high likelihood of failure. Thus, scientific knowledge seemed to have become merely a political tool to assuage the enthusiasm of a handful of naturalists and biologists for the charismatic cat, rather than the basis for a well-researched plan to implement its reintroduction in a less-than-welcoming landscape.

Entities to deliberate and include a wide range of expertise and perspectives, especially scientific, have been created in the last 30 years as the government regulatory and decision-making systems for environmental management have become more sophisticated. Expert panels such as the National Board for Wildlife (NBWL) and the Reserve Management Advisory Board are good examples of such efforts to widely consult and inform more bureaucratic mechanisms for decision-making with a variety of citizen perspectives from technical experts to grassroots activism and local knowledge. But for the most part, either these panels are not implemented, or are rendered marginal to the crucial decisions that emerge from line departments and ministries and are then confronted in courts through public interest litigation. Thus, the meetings of the NBWL sometimes fail to carefully discuss the proposed projects due to the shortage of scientists, the short time given for consideration and the pressure to clear all projects, however serious their ecological impacts might be. The meetings simply turn into a fast-track panel for clearing projects, rather than performing a considered analysis of the tricky issues and options for best policies for sustainable development or inclusive conservation (Bindra 2017). Through some of the case studies presented in this volume, these issues of availability and incorporation of scientific data into real-life management are addressed.

Economic liberalisation has, then, had impact on nature conservation in India in at least two ways. One, more obvious, is the increased stress on natural resources and habitats to provide raw materials and landscapes for development and consumption since there is a rapidly growing appetite for such things. The other, noticeable in the way conservation becomes more enterprise-oriented and more contentious around issues of rights and responsibilities, has resulted in a greater focus on ecosystem services that lend themselves to computation, prioritisation, and ultimately both regulation and monetisation. Hence, we see this project as an investigation of the ways in which the protection of natural heritage, a state-led project in the first few decades after Indian Independence, has become a realm of private enterprise, civic action and competing regulations where science, activism and government, more often than not, collide.

Notes

1. Shashank Kela, 2016, 'State Structures and Forest Laws: A Policy Perspective on Forest Conservation in India', Paper Presented at a workshop titled *Wildlife Conservation in India: From Policy to Practice*, Centre for Advanced Study of India, University of Pennsylvania, 4–5 November 2016, New Delhi.
2. Chandrima Home, and Abi T. Vanak, 2016, 'Holier than Cows? Stray Dogs as an Emerging Threat for Wildlife in India', Paper Presented at a workshop titled *Wildlife Conservation in India: From Policy to Practice*, Centre for Advanced Study of India, University of Pennsylvania, 4–5 November 2016, New Delhi.

References

Agarwal, R. 2014. 'The Fight for an Urban Forest: The Delhi Ridge'. In M. Rangarajan, M. D. Madhusudan and G. Shahabuddin (eds), *Nature Without Borders*, pp. 107–30. Delhi: Orient BlackSwan.

Amirante, D. 2012. 'Environmental Courts in Comparative Perspective: Preliminary Reflections on the National Green Tribunal of India', *Pace Environmental Law Review* 29(2): 441.

Anand, I., and A. Thampi. 2016. 'Recent Trends in Wealth Inequality in India', *Economic and Political Weekly* 51(50), 10 December 2016, pp. 59–67.

Athreya V, M. Odden, J. D. Linnell, et al. 2013.'Big Cats in our Backyards: Persistence of Large Carnivores in a Human Dominated Landscape in India', *PLOS One* 8(3). Available at e57872. doi10.1371/journal.pone.0057872 (accessed 10 March 2018).

Balakrishnan, P. 2010. *Economic Growth in India: History and Prospect*. Delhi: Oxford University Press.

Bindra, P. S. 2017. *The Vanishing: India's Wildlife Crisis*, p. 326. New Delhi: Penguin-Random House.

Carter, N. H., B. K. Shrestha, J. B. Karki, et al. 2012. 'Coexistence between Wildlife and Humans at Fine Spatial Scales', Proceedings of the National Academy of Sciences. Available at www.pnas.org/cgi/doi/10.1073/pnas.1210490109 (accessed 12 September 2017).

Cederlof, G., and K. Sivaramakrishnan. 2006. *Ecological Nationalisms: Nature, Livelihoods and Identities in South Asia*. Seattle: University of Washington Press.

Chhatre, A., and V. K. Saberwal. 2006. *Democratizing Nature: Politics, Conservation and Development in India*. New Delhi: Oxford University Press.

Das, Maitreyi Bordia, Soumya Kapoor-Mehta, Emcet Oktay Taş, et al.

2015. *Scaling the Heights: Social Inclusion and Sustainable Development in Himachal Pradesh.* Washington, D.C.: The World Bank.

Dharmakumarsinhji, R. S. 1998. *Reminiscences of Indian Wildlife.* Delhi: Oxford University Press.

Gadgil, M. 2001. *Ecological Journeys: The Science and Politics of Conservation in India.* Ranikhet: Permanent Black.

Guha, R. 2003. 'The Authoritarian Biologist and the Arrogance of Anti-Humanism: Wildlife Conservation in the Third World'. In V. K. Saberwal and M. Rangarajan (eds), *Battles Over Nature: Science and the Politics of Conservation,* pp. 139–57. New Delhi: Permanent Black.

Hecht, S. B., K. D. Morrison, and C. Padoch. 2014. *The Social Lives of Forests, Past, Present and Future of Woodland Resurgence,* p. 493. Chicago and London: University of Chicago Press.

Home, C., and Abi T. Vanak. 2016. 'Holier than Cows? Stray Dogs as an Emerging Threat for Wildlife in India', Paper Presented at a workshop titled *Wildlife Conservation in India: From Policy to Practice,* Centre for Advanced Study of India, University of Pennsylvania, 4–5 November 2016, New Delhi.

Home, C., A. T. Vanak, and Y. V. Bhatnagar. 2017. 'Canine Conundrum: Domestic Dogs as an Invasive Species and their Impacts on Wildlife in India'. *Animal Conservation.* Available at https://doi.org/10.1111/acv.12389 (accessed 13 March 2018).

Kapoor, M., K. Kohli, and M. Menon. 2009. *India's Notified Ecologically Sensitive Areas (ESAs): The Story So Far…* Delhi: Kalpavriksh; New Delhi: World Wildlife Fund-India.

Karanth, K. U. 2006. *A View from the Machan.* Delhi: Permanent Black.

Kela, Shashank. 2016. 'State Structures and Forest Laws: A Policy Perspective on Forest Conservation in India', Paper Presented at a workshop titled *Wildlife Conservation in India: From Policy to Practice,* Centre for Advanced Study of India, University of Pennsylvania, 4–5 November 2016, New Delhi.

Kothari, A. 1998. *Towards Participatory Conservation in India: National Scenario and Lessons from the Field.* Delhi: Kalpavriksh.

Krishnan, M. 2012. *Of Birds and Birdsong.* Delhi: Aleph Book Company.

Lenin, J. 2007. 'Song of the Ganges Gharial'. *Seminar* 577: 34–38.

Lewis, M. 2003. 'Cattle and Conservation at Bharatpur: A Case Study in Science and Advocacy', *Conservation and Society* 1(1): 1–22.

———. 2004. *Inventing Global Ecology: Tracking the Biodiversity Ideal in India, 1947–1997.* Columbus: Ohio University Press.

———. 2005. 'Indian Science for Indian Tigers? Conservation Biology and the Question of Cultural Values', *Journal of the History of Biology* 38(2): 185–207.

Madhusudan, M. D., and C. Mishra. 2003. 'Why Big Fierce Animals are Threatened: Conserving Large Mammals in Densely Populated

Landscapes'. In V. Saberwal and M. Rangarajan (eds), *Battles Over Nature: Science and the Politics of Conservation*. Delhi: Permanent Black.

McNeill, J. R., and P. Engelke. 2014. *The Great Acceleration: An Environmental History of the Anthropocene since 1945*. Cambridge: Harvard University Press.

Menon, M., and K. Kohli. 2015. 'Environmental Regulation in India: Moving "Forward" in the Old Direction', *Economic and Political Weekly* L(50): 20–23.

MoEF. 2005. *Joining the Dots: Report of the Tiger Task Force*. New Delhi: Government of India.

———. 2010. *Gajah: Securing the Future for Elephants in India*. The Report of the Elephant Task Force, Ministry of Environment and Forests, New Delhi.

Mohan, M. P. R., and H. Pabreja. 2016. 'Public Hearings in Environmental Clearance Process: Review of Judicial Intervention', *Economic and Political Weekly* LI(50), 10 December 2016, pp. 68–75.

Nagendra, H. 2016. *Nature in the City: Bengaluru in the Past, Present and Future*, p. 224. Delhi: Oxford University Press.

Osuri, A. M., A. Bali, A. Kumar, and J. Krishnaswamy. 2010. 'Sustaining Biodiversity Conservation in Human-Modified Landscapes in the Western Ghats: Remnant Forests Matter', *Biological Conservation* 143(10): 2363–74.

Porter-Bolland, L., E. A. Ellis, M. R. Guariguata, et al. 2012. 'Community Managed Forests and Forest Protected Areas: An Assessment of their Conservation Effectiveness Across the Tropics', *Forest Ecology and Management* 268: 6–17.

Rahmani, A. R. 2003. 'Conservation Outside Protected Areas: Case Study of Bustard Protection'. In V. Saberwal and M. Rangarajan (eds), *Battles Over Nature: Science and the Politics of Conservation*, pp. 117–38. Delhi: Permanent Black.

Ramesh, J. 2017. *Indira Gandhi: A Life in Nature*, p. 437. New Delhi: Simon & Schuster India.

Rangarajan, M. 2001. *India's Wildlife History: An Introduction*. New Delhi: Permanent Black.

———. 2015. *Nature and Nation: Essays in Environmental History*, p. 346. Ranikhet: Permanent Black; Haryana: Ashoka University.

———. 2017. *Nature's Nation, Seminar* 696, August 2017, pp. 46–52.

Rangarajan, M., and K. Sivaramakrishnan. 2013. *India's Environmental History: A Reader*, p. 1096. Delhi: Orient BlackSwan.

Rangarajan, M., M. D. Madhusudan, and G. Shahabuddin (eds). 2014. *Nature Without Borders*. New Delhi: Orient BlackSwan.

Ranjitsinh, M. K. 2017. *A Life with Wildlife: From Princely India to the Present*, p. 380. Noida, UP: HarperCollins Publishers.

Rao, M., H. Nagendra, G. Shahabuddin, et al. 2016. 'Integrating

Community-Managed Areas into Protected Area Systems: The Promise of Synergies and the Reality of Trade-Offs'. In L. Joppa, J. Robinson and J. Baillie (eds), *Protected Areas: Are they Safeguarding Biodiversity?*, pp. 169–89. West Sussex, UK: Wiley-Blackwell.

Rodgers, W. A., H. S. Panwar, and V. B. Mathur. 1988. *Wildlife Protected Area Network in India: A Review.* Dehradun: Wildlife Institute of India.

Saberwal, V. K., and M. Rangarajan. (eds). 2003. *Battles Over Nature: Science and the Politics of Conservation.* New Delhi: Permanent Black.

Saberwal, V., M. Rangarajan, and A. Kothari. 2001. 'Humans and Protected Areas'. In *Parks, People and Wildlife: Towards Coexistence*, pp. 44–70. New Delhi: Orient Longman.

Sanga, P., and A. Shaban. 2017. 'Regional Divergence and Inequalities in India', *Economic and Political Weekly*, 52(1), 7 January 2017.

Sankhala, K. S. 1977. *Tiger! The Story of the Indian Tiger.* New York: Simon and Schuster.

Saxena, N. C. 2010. 'Report: National Committee on Forest Rights Act', Joint Committee of Ministry of Environment and Forests and Ministry of Tribal Affairs, Government of India.

Shahabuddin, G. 2007. 'The Endangered Wildlife Biologist'. Available at http://www.india-seminar.com/2007/577/577_ghazala_shahabuddin.htm (accessed 10 September2017).

———. 2010. *Conservation at the Crossroads: Science, Society and the Future of India's Wildlife*, p. 288. Delhi: Permanent Black; Bengaluru: New India Foundation.

Shahabuddin, G., and M. Rangarajan (eds). 2007. *Making Conservation Work: Securing Biodiversity in this New Century*, p. 298. Delhi: Oxford University Press.

Shahabuddin, G., R. Kumar, and M. Shrivastava. 2007. 'Creation of Inviolate Space: Lives, Livelihoods and Conflict in Sariska Tiger Reserve', *Economic and Political Weekly* 42(20): 1855–62.

Shanker, K. 2016. *From Soup to Superstar: The Story of Sea Turtle Conservation along the Indian Coast*, p. 360. Noida, UP: HarperCollins.

Shrivastava, A., and A. Kothari. 2012. *Churning the Earth: The Making of Global India*, pp. 394, xxi. New Delhi: Penguin Books.

Singh, S. S., 2015. 'India's Forest Cover Up by 5871 sq.km', *The Hindu*, 15 February 2015. Available at http://www.thehindu.com/news/national/other-states/indias-forest-cover-up-by-5871-sq-km/article6896830.ece (accessed 12 September 2017).

Sivaramakrishnan, K. 2011. 'Environment, Law and Democracy in India', *The Journal of Asian Studies* 70(04): 905–28.

———. 2015. 'Ethics of Nature in Indian Environmental History', *Modern Asian Studies*, 49(4): 1261–1310.

———. 2017. 'Courts, Public Cultures of Legality, and Urban Ecological Imagination in Delhi'. In Anne Rademacher and K. Sivaramakrishnan (eds), *Cities, Towns and the Places of Nature in Asia*, pp. 137–61. Hong Kong: Hong Kong University Press.

Smetacek, P. 2012. *Butterflies on the Roof of the World: A Memoir.* Delhi: Aleph Book Company.

Springate-Baginski, O., and P. Blaikie (eds). 2007. *Forests People and Power: The Political Ecology of Reform in South Asia.* London: Earthscan.

Sundar, N., R. Jeffery, and Neil Thin (eds). 2001. *Branching Out: Joint Forest Management in India.* Delhi: Oxford University Press.

Thapar, V. (ed.). 2001. *Saving Wild Tigers 1900–2000*, p. 411. Delhi: Permanent Black.

TEEB. 2010. *The Economics of Ecosystems and Biodiversity: Mainstreaming the Economics of Nature: A Synthesis of the Approach, Conclusions and Recommendations of TEEB.* Available at http://www.teebweb.org/our-publications/teeb-study-reports/synthesis-report/ (accessed 26 December 2018).

WII. 2017. 'Protected Areas of India'. Available at http://www.wiienvis.nic.in/Database/Protected_Area_854.aspx (accessed 10 September 2017).

1

Hunting in Northeast India and the Challenges of Implementing the Wildlife Protection Act

Ambika Aiyadurai

Hunting of wild animals is a criminal offence according to India's Wildlife Protection Act, but it is a culturally acceptable practice among several indigenous communities. After close to five decades, implementing this Act to curb wildlife hunting has been difficult, especially in northeast India. While the degree of hunting may vary across the region depending on factors like distance from urban centres, education levels, availability of alternate livelihoods, the presence of non-governmental organisations (NGOs) and the forest department, wildlife hunting is widely practised in northeast India. In addition to cultural compulsions, other reasons for hunting in this region include crop protection, to secure additional income and as a leisure activity. In this chapter, I examine the reasons why the implementation of a comprehensive wildlife policy such as the Indian Wildlife Protection Act has been a major challenge in northeast India. I make two arguments: (1) the cultural and livelihood issues are linked to hunting and that undermines restrictions imposed by the government, (2) hunting practices are invariably linked to the dynamic realities of making identities and lives in the region. Using field observations in Arunachal Pradesh, I shall demonstrate why hunting is likely to remain and possibly take different forms and degrees to adapt to rapid changes in the socio-economic and cultural fabric of indigenous societies in northeast India.

Introduction

I was in a meeting with a Delhi-based NGO, the Arunachal Pradesh Forest Department and members of a local community in Anini (Dibang Valley, Arunachal Pradesh). The objective of the NGO was to map the Dibang Wildlife Sanctuary. The discussion during the meeting revealed a wide gap in the way different groups understand the practice of hunting and conservation. The talk of tigers is inevitable whenever there is a discussion on forests in this region. For the last three years, there have been surveys by wildlife biologists and NGOs to estimate the number of tigers. During the meeting, Anand, a trained wildlife biologist, explained to the members of the local community the importance of tigers as top predators in the ecosystem and how tigers keep a check on the herbivore population. He patiently explained the role of tigers and exhorted the villagers, 'If tigers are killed, the prey population will increase and destroy the forests.' This textbook-based food chain concept was demolished in a second when Athuko, a Zilla Parishad member, replied casually and confidently '... if the deer population increases, we are there to hunt the deer and we will keep the deer population under control.' Anand and his NGO team were taken aback since, in fact, hunting of wild animals is forbidden. Athuko qualified his statement with pride and smiled, '... because we don't kill tigers, so the deer population is controlled by us.' Athuko grinned, followed by laughter. I have witnessed such discussion several times, where hunting of wild animals is discussed openly during meetings, where the local community takes pride in their hunting practice even in the presence of forest officials. This reveals a critical difference between the views of the local people and the official authorities and wildlife biologists regarding hunting. In this chapter, I write about the Mishmi (the Idu Mishmi), and largely about the rural Mishmi's way of life and their relation with animals and spirits to demonstrate why this law has failed in places like Arunachal Pradesh.

Different tribes across northeast India openly speak of hunting. Conversation about skulls and skins of wild animals displayed in people's houses, sometimes neatly mounted on bamboo frames, were easy. Display of skulls is a fairly common practice even today. For example, in Miju Mishmi areas in eastern Arunachal Pradesh, houses and porches are decorated with the mounted skulls of prey like the Indian muntjac *Muntiacus muntjak*, the wild pig *Sus scrofa*, the takin *Budorcas taxicolor* and the Asiatic black bear *Ursus thibetanus* (Aiyadurai 2009). Also on display among the Mishmi are fans made from tail feathers of the pheasant or the eagle, head gear from hornbill casques, machete sheaths and shoulder belts made from the capped langur *Trachypithecus pileatus*, and bear skins that are commonly used by the Nyishi people (Aiyadurai 2007b). Similar observations were made on the use of skulls on 'trophy boards' across the border in Myanmar (Rabinowitz 2001).

In this chapter, I focus on the examples from Arunachal Pradesh, largely from the Mishmi hills, a region I am most familiar with. At the outset, I wish to state that I would not like to speak of all 'tribal'[3] groups or all members of one particular tribal group as one entity because they are not. People living in towns have different social positions, and hence, there are many different views from various members within a particular indigenous group, viz., urban and hill dwelling, elite and non-elite, males and females. I aim to present the world of the Mishmi that is complex and its culture as not stable or intact but as changing with time.

Background

The Indian Wildlife Protection Act, 1972 (Government of India [GoI] 1994) prohibits hunting[4] of virtually all large wildlife. The main objectives of the Act are: (1) prohibition of hunting of wild animals, birds and plants, (2) setting up and management of national parks and sanctuaries, (3) control of trade and commerce in wildlife and wildlife products. The Indian Wildlife Protection Act is a comprehensive law for protecting India's biodiversity and it prohibits hunting of any species, as well as trade in trophies, animal parts and their derivatives. A complete ban on hunting of wildlife species has affected the livelihood of many communities in India (Dutt 2004; Dutt et al. 2007; Gadgil and Malhotra 1998). The law was created in response to the rapid decline in India's wildlife and international pressure on forests and wildlife conservation. With the creation of these protected areas (PAs), the subsistence activities of the local people were deemed illegal; for example, according to the Act:

> No person shall destroy, exploit, or remove any wildlife from a National Park or destroy or damage the habitat or any wild

> animal[5] or deprive any wild animal or its habitat within such National Park except under and in accordance with a permit granted by the Chief Wildlife Warden and no such permit shall be granted unless the State Government .... authorizes the issue of such permit.

The Act further states: 'No grazing of any livestock shall be permitted in a National Park and no livestock shall be allowed to enter except where such livestock is used as a vehicle by a person authorized to enter such National Park.'

'Our laws are such of a kind that every villager breaks one forest law every day of his life', said Verrier Elwin (1964: 115). A well-known anthropologist who lived and worked with Indian tribals in the mid-twentieth century highlights how a law such as the Wildlife Protection Act has impacted the local users of natural resources and their subsistence activities. For many indigenous people, carrying out farm-related activities, trade, or certain religious or social events, are contradictory to the Wildlife Protection Act, which makes them 'ecological criminals'.

India's Wildlife Protection Act was enacted in response to the rapid decline in wildlife populations due to hunting and habitat modification. Several studies indicate that the ecological implications are severe. Hunting for, and trade in, animal parts is a major threat to wildlife across the tropics (Bennett et al. 2000; Milner-Gulland et al. 2003). With growing human populations, increased accessibility to remote forests and the adoption of modern hunting methods and guns, the problem of hunting has increased in many parts of northeast India (Aiyadurai et al. 2010; Datta et al. 2008; Velho et al. 2012).

Arunachal Pradesh poses certain challenges in terms of logistics, climatic conditions and cultural milieu that are vastly different from mainland India. For example, Dibang Valley district in Arunachal Pradesh has the lowest human density in India (1 person per sq. km). The local Mishmi population is largely dependent on land for swidden farming, and they have a difficult life, as the land is largely unsuitable for any kind of agriculture, especially on the higher mountain ranges. In the last couple of decades, the Government of India has been investing in developing this area, which has been rapidly shaping the region.

Mishmi Social Worlds

Like many tribals, the Mishmis have both pragmatic relations and a cosmological relationship with animals. For all of them, nature is 'one', including the world of 'spirits'. The Mishmis constantly engage with animals and spirits during farming, hunting wildlife or slaughtering domesticated animals. Scholars have highlighted that social attitudes to and relationships with animals are never monolithic and all human societies have complex, diverse, multifaceted ideas about animals (Bird-David 1990; Morris 1998). Brian Morris, for example, highlights attitudes towards animals that are not only intuitive, psychological and imaginary, but he also shows how people's attitudes towards some animals may be pragmatic at the same time as framing the relationship in terms of close kin relationships (Morris 2000). Some animals may be conceived to be under the control of the guardian or forest spirits, who are themselves interconnected with the social world. The Mishmis[6] believe that spirits take care of domesticated animals (*Mitu-sipa*) and wild animals (*Golon*); spirits are often talked about as if they are people too, and this belief is so strong that it means respecting and being fearful of the spirits. For example, when asked why women are prohibited from consuming wild meat, an old Mishmi lady told me '*Golon humko maarega*' ('*Golon* will beat us'), and that *Golon* watches people from afar. Therefore, women are prohibited from consuming wild meat and reference to spirits is very common.

The Mishmi's social world is a network of their association with humans, animals and spirits. The presence of spirits in farms, homes, forests and mountains, and the significant role of these spirits, is reflected during harvesting, healing rites, funerals and birth ceremonies. Vidyarthi (1963) proposed the concept of 'nature–man–spirit' complex. In his research among the Maler tribe of Bihar (India), he showed that the Maler is part of a world where three components (nature, humans and spirits) are independent and mutually complementary. This, according to him, is an interaction of necessity arising from the fact that the Maler interacting closely with nature is part of both religious and social organisations, similar to the Mishmis.

Hunting among the Mishmis[7] is part of an ancestral way of life and a cultural practice. Men undertake hunting for subsistence, trade[8] and crop protection. Villagers lay different types of traps around the farm boundary to protect the crops. Maize (*ambo*) and millet (*yamba*) are the major crops grown in these plots, along with a variety of vegetables (*anapra*). Bears (*ahun*), wild boar (*aamme*) and monkeys (*aame*) are reported to raid crops during the harvest period (June–July). Long-distance hunting trips, especially ones that are undertaken for musk deer (*ala*), takin (*aakru*) and other ungulates, are often carried out in the mountains. Hunters in Dibang Valley reach close to the Sino-Indian border for musk deer hunting. Animals such as barking deer, wild boar, monkeys and bears are commonly hunted closer to the villages or when they raid crops, especially during harvesting maize. Hunting is carried out alone, in small groups or with trusted partners[9] (other village men).

Some examples of the role of hunting and wild meat in the local communities include offering of wild meat as bride price during weddings among the Miju Mishmi and it is regarded as a status symbol (Aiyadurai 2009). Among the Idus, men collectively go for hunting to gather enough wildmeat for the weddings. Orange-bellied squirrels (*Dremonys lokriah*) are gifted by Adi men to the bride's family during wedding ceremonies (Das and Shukla 2007). Wildlife products are exchanged to gain goodwill and to establish or affirm relationships of patronage, even in fast-changing urban or semi-rural towns. The gifting of wildlife products is also a common way to gain goodwill from senior government officers and elite members of society. When visiting, government officers are, to this day, presented with animal skins that are used as decorative items at their residences.

Dibang Valley district is known for the Himalayan musk deer (*aala*). Musk pods are the scent glands of male musk deer, which are in demand in international markets because of their use as perfume. One of the hunters said that this animal carries 'gold', by which he meant musk pods, which has a high market value, both in local and in international markets. The cost of a pod is around Rs 20,000–30,000 ($300–450) per *tola.*[10] Meat is consumed and musk pods (*aalaapee*) are sold quickly. The pods are not used locally but are often sold to Marwaris,[11] which are then exported illegally to international markets and used in making perfumes. Another animal that is hunted for trade-related purposes is the black bear. It is targetted for its gall bladder, from which bile is extracted and used in traditional Chinese medicine. It is sold locally for Rs 5,000 ($75) per *tola.* The demand for soft and waterproof otter skins is high in the international trade, and each skin is sold for Rs 8,000–10,000 ($120–50) to buyers from outside the region, sometimes even from Myanmar.[12] Markets create new opportunities, mobilise skills, generate income and enable new forms of accumulation and a new level of pressure on resources.

Hunters and Animal Guardian Spirits

The Mishmis do not take long-distance hunting (*anji*) casually (Aiyadurai 2016; Blench 2016). Hunters (*anjiya*) equate hunting to a spiritual activity with many taboos (*angein*) that need to be observed carefully. Taboos are observed before, during and after hunting. *Aambo*, a type of generic ritual that is performed to ask for success and safety in any endeavour, is performed by an *igu* at a hunter's house before the trip. During this ritual, hunters pray for hunting success and for their own safety. An *igu* slaughters a chicken and its blood is offered to the spirits. In exchange for this offering, the *igu* asks for success in hunting. Hunters leave very early in the morning before sunrise. Once they reach the hunting grounds, a symbolic gate (*la-ro*) is prepared towards the direction of their movement. A short prayer is chanted by one of the hunters before crossing this gate. The gate is believed to be a symbolic boundary between the village and the hunting zone. Crossing the gate actually indicates the beginning of hunting, which requires the hunters to follow certain rules. This is seen as the beginning of a 'spiritual' journey during which hunters engage with forest spirits and animals. This is a kind of 'rite of passage' where the transition is made to a spiritual world and where the Mishmis enter into a 'social contract' with animals and spirits. In the Mishmi worldview, there are certain duties, responsibilities and rights that hunters have towards animals and spirits, founded on terms of engagement that depend on trust, reciprocity and mutual assistance. Hunters symbolically share food and drinks with the spirits (a piece of meat or a leaf cup filled with rice beer, *yiu*) as one would share with other people. Both animals and spirits are seen as socially and spiritually equivalent and are believed to have feelings like people. Therefore, rituals are performed for animals in the way people perform rituals in their homes. Rites always accompany hunting trips as a mark of respect and reverence to animals. Offerings are made with the expectation that the spirits will be pleased and, in turn, they will provide more animals in the subsequent trips.

Spirits are considered to be active agents and wild animals in this hunting zone are believed to be owned by a mountain spirit, *Golon. Golon* is the most important spirit for hunters and is believed to live in the high mountains and thick forests. People believe that *Golon* is the one who 'supplies' animals to the hunters, and thus, when there is a successful hunt, an offering has to be made to *Golon.* If not, *Golon* gets angry and could cause accidents to both hunters and consumers of the hunted animal, and may cause severe illness to either the hunter or his family members. Hunters have duties, responsibilities and rights, which need to be followed to let *Golon* offer animals to hunters. One of the duties of hunters is to address the animals in a proper way with special code names. Wild animals have two names: one is the Mishmi name that is used when they are in the village, and the other is the name that hunters know and use only during hunting, an exclusive knowledge that only hunters have.

After a successful hunt, the hunter performs a ritual (*aanphun aangi*). This ritual symbolises making a payment in the form of meat and a piece of metal to the spirit, *Golon.* ' ... *Hum ko daam dena padta hai*', said a hunter, which translates to ' ... We have to make a payment or pay a price.' A small piece from the ear of the hunted animal, inserted into a bamboo fork, is planted on a nearby tree or on the ground. Using a *dao*, the hunter scrapes the brass metal (*aanphoodi/paao yute/seekhi yute*) over the dead animal and recites a chant, which roughly translates as:

> This animal has been killed. When I come again, please give more animals. One who rears the animal, let more animals be produced. Give us more animals whenever we come again. We have given you the payment, do not disturb or harass the one who eats this meat and the one who hunted this animal. The restrictions will be observed and do not follow us and watch us whether we did the right thing or not. Do not disturb us.

When the hunter returns home, there is a restriction on the things the hunter can or cannot do for five days, but only if it was a successful hunt. There are taboos on eating onion (*elompran*), garlic (*eloni*), mushroom (*akupi*), a herb (*marsana)* and fermented soyabean (*aadulo chhin*). Men stay in a separate room and sex is forbidden. They are not allowed to receive water and tea prepared by women who are menstruating. Hunters avoid attending funerals and washing clothes. The animal skulls are mounted on the wall of the men's rooms. Displays of wild and domesticated animal skulls are a common sight among the Mishmi. The wild boar, takin, serow, goral and Asiatic black bear are some of the animals whose skulls are mounted on a neatly made bamboo frame, which is found in almost every rural Mishmi house. Animal skulls and jaws of animals are symbols of hunting and sacrificial rites. Skulls remind hunters of special events related to hunting or slaughtering of animals. The relationship between animals and male identity is prominent among the Mishmis, and this reflects in the way the men maintain both wild animal skulls and domestic animal skulls displayed on the trophy board.

Taboos for the Protection of Wild Animals

While hunting of some animals is permitted provided one follows the taboos, there is a restriction of hunting of animals such as tigers and the hoolock gibbon. Killing a tiger is the worst crime a man can commit and a strict taboo is observed not only by the hunter and by his family members, but also by the entire village. Anyone visiting the Mishmi hills and interested in wildlife and conservation cannot fail to encounter the mythological story of the Mishmis and tigers as brothers. Such narratives of tigers as siblings are popular in other parts of Arunachal Pradesh (Aisher 2005; Aiyadurai 2007a). According to Mishmi mythology, the Mishmis and tigers were born to the same mother and were siblings (tiger, the elder brother and human, the younger brother). The younger brother hunted a deer and left it with the elder brother to collect firewood. On his return, he was terrified to see his brother eating the meat raw. He told his mother, 'My younger brother is a tiger. If he can eat the raw meat, then one day he will eat me too.' The mother came up with a plan: she would hold a competition between the two brothers. The one who crossed the river and was first to reach the bank would kill the other. The tiger decided to swim across the river, whereas the Mishmi took the bridge. The tiger was the first to reach the bank, but as he emerged from the water the mother threw an ant-nest on his body to prevent him from winning. The tiger dived back into the water and scratched his body against a rock. The Mishmi, meanwhile, reached and climbed up the bank and shot the tiger with an arrow. Thus, the tiger died and its body floated in the river and was swept away to a faraway place. Several years later, a bird saw the bones of the tiger scattered on the riverside. The bones were white and shone bright in the sunlight. Thinking these were eggs, the bird sat upon them. From these bones came the tiger, and from the smaller bones came the leopard, leopard cat, clouded leopard and civet cat. This is the story of a tiger being born again. This story is also why the Mishmis refrain from killing tigers. For them,[13] the tiger (*aamra*) is their elder brother (*apiya*). It is the most revered and feared animal, and killing it is prohibited. In fact, it is viewed as 'homicide'.

According to Morris (2004), treating animals as persons sets up a relational epistemology towards other animals and other 'natural' beings. In the case of the Mishmis, tigers are conceived as their brothers. Therefore, as a rule, they never kill or trap them, unless tigers become a threat to human lives or are accidentally trapped. It is important to distinguish carefully between social practices and cultural representations. If killed or trapped accidentally, a senior *igu* will be invited to carry out the required ritual (*taamaamran*), which involves a huge expenditure because the ritual is equivalent to a funerary ritual conducted for humans. During this elaborate ritual carried out over five days, strict taboos (*angein*) are observed not only by the person who killed the tiger, but also by the entire village. Every *igu* owns a belt (*amrala*) made of linearly arranged tiger canines, through which he receives power. This belt is a necessary ritual item for *igus* while performing a funeral or any other important rituals like *amrase*[14] or *Reh.*[15] To prevent attacks by spirits, *kheenu*, *Igus* wear tiger tooth belts to awaken the spirit of the tiger.

Human–Animal Relations

The Mishmi's relation with animals ranges from protecting some animals (for example, tigers and the hoolock gibbon), utilising some animals for meat (takin, serow, barking deer), trading animal parts (pods of musk deer and gall bladders of bears), and if required, killing carnivores that attack their cattle (wild dogs, tigers) or animals that raid their crops (bears, ungulates, wild pigs). While killing animals like tigers and gibbons are conceived as equivalent to homicide, other animals like the takin *Budorcas taxicolor*, musk deer *Moschus chryogaster*, Himalayan black bear *Ursus thibetanus*, flying squirrels and pheasants are no less frequently hunted. The use of these animals is diverse and widespread: meat as food, skins as bags, and mats and skulls as sacred objects to be mounted on skull boards, *amuneenddon*.[16] Therefore, many do not see hunting as a harmful activity. It is seen as an activity that protects their livelihood, human lives and as a culturally just practice to maintain their social network. Certain rules and taboos help in conserving animals, and therefore, not protecting crops and not hunting animals is considered as illogical.

A discussion of types of 'person' is useful here. Different kinds of animals are treated differently depending on the degrees of personhood. Some animals are completely protected as other human beings, while others are seen as non-human animals with moral agency and consciousness. On the same lines, Tsing (2013) argues for a more-than-human-sociality through which she advises scholars to explore 'multispecies landscapes' and dynamic relations between different species. These relations 'create web of social relations and not just individual enrollment for human tools' (ibid.: 27–42).

Local people take pride in wildlife hunting and their indigenous taboos become a sort of 'cultural capital' (Bourdieu 1986). The Mishmis often claim that they hunt with 'responsibility' and their indigenous taboos on hunting are better and stricter than those of the other groups, and therefore, they contribute to the preservation of local wildlife. This becomes a point of contention with biologists and NGOs. There is a sense of identity that the people of Arunachal Pradesh take pride in, regarding hunting practice and in terms of material culture: Nyshis wearing hornbill beak casque as a head gear, carrying a bear bag among the Mishmis, or wearing hornbill feathers as a head dress among the Wanchos. Local people in Arunachal Pradesh carry out hunting of takins and they distribute the meat to the villagers, a form of hunting that is risky but with little return, except the meat. Other forms of display of animal remains can be found in houses: some bones haphazardly accumulated in a corner or systematically arranged on the skull board.

The pioneering work of Alex Aisher with the Nyishi community of Arunachal Pradesh throws light on the forest spirits and how they influence the way people feel about their interactions with the ecosystem, thus emphasising the intricate relationship local people share with their environment (Aisher 2007). Among Asian hunting societies in China and Indonesia, hunting is a significant social activity and wild animals are powerful as subjects of myths, symbols and omens (Donovan 2004; Knight 2004). Hunting continues to play an integral part in several indigenous societies in spite of major social and economic adaptations and remains at the 'heart of the culture' as described for Agta indigenous groups in the Philippines (Griffin and Griffin 2000).

Among the Mishmis, there is a continuum of the animal/human world dealing with the variable nature of the Mishmi's adherence to cosmological principles. In their cosmos, they relate different animals to humans as ancestors, spirits and co-residents of a sacred landscape. In countries such as Japan and Sweden, these 'spirit-imbibed landscapes' are believed to be an actual embodiment of spirits and gods who inhabit these sacred sites (Byrne 2010). In Sweden, Saami peoples involved in reindeer hunting see their land as a 'living entity' and people have a special relationship with the landscape. According to them, these landscapes are sacred and are inhabited by deities, ancestors and spirits (Broadbent and Edvingen 2011).

Conserving Arunachal's Wildlife

There has been a serious concern globally over wild animal extraction and research that indicates that hunting is leading to the extinction of wildlife populations (Bennett and Robinson 2000). Northeast India has recently been subjected to a growing academic interest among ecologists and conservationists. Wildlife research increased in the region, partly, because of the inclusion of the region within the Eastern Himalaya 'biodiversity hotspot' (Myers et al. 2000).

Setting up of PAs has been one of the major approaches to conserve wildlife. Arunachal's Forest Department has set aside large tracts of land for conservation as national parks and wildlife sanctuaries. There are eight wildlife sanctuaries and two national parks, which are also designated as tiger reserves in the state. The Forest Department's engagement with the local communities for wildlife conservation has been slow and for a long period, this was not a priority for the government. NGOs (World Wide Fund for Nature, Wildlife Trust of India [WTI], Nature Conservation Foundation) have taken an active role in conservation in PAs, viz., in Pakke Tiger Reserve and Namdapha Tiger Reserve. In the remote border areas, Dibang Wildlife Sanctuary could be considered to be what is termed as a 'paper park'. This term is applied to PAs that exist in an 'administrative vacuum' (del Valle 2002: 150) without proper maps, boundary demarcation and with insufficient field staff. With the increasing focus on tigers in Dibang and the possibility of declaring Dibang Wildlife Sanctuary as Dibang Tiger Reserve, there is a feeling that local people can do better in conserving wildlife than the Forest Department. Government officials also consider their postings in Dibang Valley as 'punishment posting' due to its remote location close to the Sino-India border. Dibang Wildlife Sanctuary is spread over half the Dibang Valley district. With just seven staff to manage 4,914 sq. km, it is extremely challenging for the department to patrol and carry out any surveys or studies.

As the local people take pride in their local taboos, complete restriction of hunting is seen as illogical. There is a real concern that when the State takes interest in tiger conservation, there will be more and stricter regulation of people's access to wild animals. There is a feeling among many Mishmi people that resources are better conserved if customary laws are in place. Many express concerns that increased government regulation will result in a diminished sense of ownership and care for local resources, and end up with local people over-exploiting such resources. One of the residents of Dibang Valley said,

> See this jungle is my clan's. We save the animals here. We can go whenever we want, without permission, but for hunting, we have to take permission from the clan member. If as a clan member, I do not permit, then he cannot hunt. If this becomes the government property, then any one can hunt, it's free will.

These anxieties also manifest in disagreements about the protected laws and conservation programmes. The staff of the Forest Department is looked down upon by the local Mishmis; because the officers and staff are from outside, unfamiliar with the terrain, trails and lack of local ecological knowledge make these officers at times vulnerable in front of the local villagers, as evident in the discussions between Anand and Athuko.

Another level of complexity is the geopolitical location of the states in northeast India. This region of Arunachal is strategically important for both China and India. It was only after the 1962 war that India began building infrastructure in northeast India and the region entered a new nationalist discourse aimed at 'nationalisation of the frontiers' (Baruah 2003: 917). This process has made regions such as northeast India financially dependent on the central government. Since then, there has been tremendous demographic and socio-cultural change with 'significant social, environmental and political costs' (ibid.). However, the development of Arunachal and northeast India in general has been shaped by a concern for national security, and therefore, Baruah argued that it is only in the cosmetic sense that Arunachal has witnessed development. The geopolitical location and the rich forest-based resources makes this landscape an internationally contested region where the focus is on national security, nature conservation and development where the military, dams and corporates play an active role.

Conservation laws and NGOs consider local people to be harmful to the wildlife species in their areas, and natural resource extraction by local people to be unsustainable, resulting in the decline in forest cover and wildlife populations (Datta et al. 2008; Hilaluddin et al. 2005). Community-based conservation projects have been implemented in the state to wean hunters away from hunting, as well as awareness programmes and the setting up of schools and health programmes to encourage local communities to participate in conservation programmes (Datta 2007). At the conceptual level, contradictions can be seen from the differences in the way nature/animals are perceived by the local communities and the law. It is really difficult to convince people who live close to the wild animals not to hunt or co-exist with nature. The only convincing argument one is left with is that 'don't hunt, because it is illegal' or 'follow the rules because the law says so.' Convincing local people not to hunt requires an enormous amount of effort and persuasion given the fact that some local people do not see extinction as a matter of concern and deny that extinction is a possible scenario.

Efforts in creating awareness, reaching out and providing alternate options are being attempted in northeast India. Amendments in the Wildlife Protection Act, 1972, to create avenues for inclusion of local communities, such as community reserves and conservation reserves are seen as some of the possible ways to take the dialogue forward. The bigger challenge is the perceptions of local people on extinctions. Most villagers perceive wildlife as an inexhaustible resource and do not expect wildlife to go extinct due to hunting (Aiyadurai, pers. obs.). The perception that wildlife is in plenty is strongly rooted, and therefore, local people may not get convinced to actively participate in conservation projects.

The role of local communities, civil societies and biologists in preserving wildlife population is showing results in the field. For example, in the case of the Amur falcon in Nagaland, people decided to stop hunting and instead use the presence of falcons in tourism-based livelihood and to attract researchers. Some NGOs and naturalists took it forward in consultation with the local community. Such a reconciliatory middle-path to combine the interests of the local people and tourism could be seen as a way forward (Kashyap 2016). People of three villages of Nagaland have made a pledge to protect and conserve the migratory Amur falcons, instead of hunting for the birds' meat (Press Trust of India [PTI] 2013). A resolution was passed by the villages to penalise offenders up to Rs 5,000 ($75) from 2013. The state government also took up the issue very strongly and the Forest Department was instrumental in enforcing the law. Within the Forest Department, officers have been put on special duty to ensure that the Amur falcons are not illegally hunted and the community remains involved (Sinha 2014).

In Pakke Tiger Reserve, the Forest Department has key leaders from the local community and works closely with village councils, conservation and research programmes, civil society, and members from the general public. Nyishis living near this PA work closely with NGOs to protect hornbill nests and prevent hunting of hornbills. There is a fine imposed on hunting of hornbills and cutting of nest trees. Fibreglass hornbill beaks in their headgear were replaced by fibreglass beaks (Kumar and Riba 2015). *Gaon buddas* (village heads) are paid an honorarium to enforce their customary conservation laws on hornbills and other species. Pakke Tiger Reserve has become an example of the multifaceted efforts required to conserve forests and wildlife with enormous dedication from NGOs and the Forest Department, but the social impacts of these efforts are yet to be studied, the political dynamics are always in flux and the sustainability of the programme is a concern.

In Namdapha Tiger Reserve, community-based conservation projects have been attempted with little or no forest department presence on the ground. The local Lisu were employed as field assistants, data recorders and guides; this was aimed at providing alternative livelihoods for the community in order to wean them away from hunting (Datta 2007). People were made to pledge not to hunt wild animals; this, however, did not succeed due to the lack of alternate sustainable livelihood options and local institutional support (Datta 2015).[17] It is still unclear whether civic and social welfare projects are effective in developing a good rapport with the community as well as in achieving conservation goals. The sociological impacts of conservation remain untested (see reports in 2012 where residents fired at members of civil society and the forest department of Namdapha Tiger Reserve). Involving local residents in conservation through community development projects may have good intentions, but they also require engaging with the socio-economic, cultural, political and historical factors that affect the use of natural resources. There have been no sociological studies of these community-based conservation projects to examine their impact on the ground.

Conclusion

This case of Mishmi hunting demonstrates a fundamental clash of two worldviews: hunting as a driver of wildlife decline and hunting as integral to the socio-cultural milieu and identities. Another important clash of worldviews is how nature is seen as a separate entity to be protected and controlled, and another, as a continuum of the nature/society or human/animal binaries. Finding a way forward without compromising the needs of rural communities requires a much greater understanding of ecology, anthropology and the socio-economics of hunting. We also need to examine how biologists, conservationists and foresters understand the issue of hunting. In my earlier work (Aiyadurai 2011), I have highlighted the need for an interdisciplinary understanding of hunting. Many of my wildlife biologist colleagues have been critical of studies that mention hunting as a cultural practice. 'Why do you call it a practice? Is it not illegal to hunt?', asked a biologist in a place where legality is seen as a basis of an ecologically grounded worldview. In documenting hunting as a 'practice' or 'custom', 'culture' is seen as 'legitimising' the act of hunting. To understand why hunting continues in the northeast, one has to look beyond the legal aspects, emotions, and animal or human rights issues and see this issue beyond one's own disciplinary lens. In this chapter, I have tried to provide insights from within an indigenous group in the northeast on how hunting is intricately woven with human lives. There are similarities across several tribes in the region and across Southeast Asia (Howell 1984; Knight 2004). In the northeast, the issue is also part of a larger disconnect due to reterritorialisation. In the post-North-East Frontier Agency (NEFA) phase, new kinds of laws, in contrast with the customary laws that exist and are followed in the region (Chaudhuri 2008; Elwin 1964).

Wildlife hunting is likely to continue in the northeast and it would possibly take different forms and degrees to adapt to rapid changes in the socio-economic structure of the rural societies in northeast India. Demand for land for constructing buildings, roads and areas for settlement would encroach upon forests. Hunting practices, in the future, may become an insignificant economic activity but would continue to be a symbolic and a ritualistic one. The enforcement of wildlife laws is likely to increase, but hunting would continue to function as a high-risk activity for those who practice it. As hunting is culturally linked, the possibility of conflicts between the government and villagers is likely in the future, especially when enforcement spreads to remote parts.

The cultural fabric has, over the past few decades, become subjected to further change with the arrival of Christianity and increasing influences of Hinduism. For example, religious sects of Hinduism such as the Art of Living (AoL) and the Brahma Kumaris promote vegetarianism and teetotalism. Although some may see this as a positive approach as such, these influences may lead to the giving up of hunting. The preaching by these newly introduced religions (both Christianity and Hinduism) undermines traditional belief systems and overlooks the already existing practices embedded in people's daily lives. Species that were previously protected by tribal hunting taboos may no longer enjoy protection under Christian mores. Interestingly, while spirits may no longer be worshipped and skulls may no longer occupy pride of place in the house, hunting still continues (Aiyadurai 2011).

Other factors like the increasing presence of mainstream Hindu religious groups and their influence on local cultural beliefs cannot be ignored. Roing town has numerous Hindu temples built by the people who come from mainland India. The mainlanders, largely Hindu-caste groups, look down on the Mishmis and other 'tribal' groups. There is a subtle undercurrent of their influence on the Mishmis, some of whom are abandoning slaughtering rituals, which has even resulted in some Mishmis becoming vegetarians. Some Mishmi members who follow the Brahma Kumaris[18] have given up meat, fish and even eggs.

During my fieldwork, I met a lady from Bangalore, a member of the AoL[19] who arrived in Roing to teach yoga and spirituality to the local Mishmi students. Members of the AoL and the Brahma Kumaris carry out programmes with school children and I have heard them preaching their views on eating habits. Visits by such groups often promote vegetarianism, indicating that beef-eating is a crime and a sin. The lady from AoL told Mishmi teachers and students who came to learn yoga to give up onions, garlic, tea/coffee and alcohol, and of course, meat! These initiatives gel well with the aims of the Arunachal Vikas Parishad,[20] a mentor of the Idu Mishmi Cultural and Literary Society (IMCLS), who are attempting to merge animistic Mishmi beliefs with Hinduism.

There is a temple in Roing, which has images of both Hindu and Mishmi deities; the temple is named Adiju-Shiv *mandir*. *Mandir* is the Hindi word for temple. *Adiju* is the name of a spirit in Mishmi mythology. Similarly, there is a temple in Hayuliong (Anjaw) called Mishmi-Ram *mandir*. Such hybridisation may lead to a gradual fusion of animism with Hinduism, marginalising or undermining the Mishmi's way of life. Such preaching by socio-spiritual groups is increasing in the remote areas of Arunachal; they reproduce the perception of the Mishmi's cultural practices as primitive, uncivilised, wild and *junglee*[21] and the need to reform such practices. Hinduism may therefore also be influencing their decision to give up the *mithun* as a symbol of their cultural and ecological identity, and this is not just an influence of global conservation. While the Mishmi take pride in 'being tribal' and in their indigenous conservation taboos, they also resist labelling their culture as 'barbaric'. Therefore, to resist this, some of them are abandoning eating animal meat and are inculcating dominant Hindu practices. The impact and influence of religious conversions on indigenous belief systems related to wildlife conservation remains to be investigated and revisited in depth.

The mismatch between the perceptions and expectations of the conservationists and the local communities is crucial. For the local residents, people hunt animals and see themselves embedded in the ecosystem, not absent or radically separate from it, unlike forest officials and scientists who see the ecosystem as 'natural', and thus, the need to promote preservation and delimit interference in the ecosystem. The local Mishmi view tiger and deer populations as a more-than-human flourishing, and not just a purified 'nature' that needs to be protected from humans. It is due to this difference in perception, attitude and relation with animals that hunting animals is largely 'not seen as a crime' or an act that is harmful for the ecosystem, as prescribed in the Wildlife Protection Act. I have argued that the reason why wildlife policy in northeast India has not been successful is because the practice of hunting is embedded in the culture of northeast India's indigenous population, linked to their identities and lives, and this is an important reason why the law has failed and is unlikely to succeed completely.

When laws such as the Wildlife Protection Act do not work in northeast India, what options do we have? This needs a debate and open discussion involving various partners including the local community and their views on hunting laws. In the northeast, hunting is a way of life, a livelihood issue, a sociocultural dimension of living, part of a trade network, or all these put together in varying degrees of complexity. Identifying the causes (and not the symptoms) and integrating them in a multi-tiered framework of governance would benefit conservation.

Notes

1. Names in this chapter have been changed to maintain anonymity of the respondents.
2. District Council.
3. I use the word 'tribal' consciously due to the complex meanings attached to it.
4. Hunting in the Wildlife Protection Act, means the capturing, poisoning, snaring or trapping of any wild animal.
5. Wildlife in the Wildlife Protection Act refers to non-domesticated animal species, including all plants, fungi, and other organisms that grow or live wild in an area without being introduced by

humans. 'Animal' includes mammals, birds, reptiles, and their young ones, and in case of birds and reptiles, their eggs as well.

6. Most of the information in this chapter comes from the Mishmis living in Anini and nearby villages. The Mishmi belief in spirits is common, but mostly, taboos are followed strictly by the Mishmis living in the villages and not in Roing.
7. Not all Mishmis. The Mishmis living in Anini mainly go for hunting on a regular basis. My information on hunting among the Mishmis comes from the men who live in Anini and villages beyond that. People in Roing often suggested that I speak to people in Anini about animals and spirits as they are 'culturally rooted' compared to people in Roing. Hunting can impact population even if at a low intensity. One has to be careful not to romanticise hunting. Early mammals' extinctions in Wrangel Island in Siberia is an evidence of the impact of low intensity hunting.
8. Species like musk deer and black bear are hunted for the musk deer pod and gall bladder, respectively. These are in demand in the international market for their use in traditional Chinese medicine.
9. Some hunters develop trusted partnership with other hunters and tend to carry out trips only with them. For example, a resident of Dibang Valley and a medical doctor (earlier posted in Anini, and now based in Roing) made several successful trips together. Compatible working relations and mutual trust help in planning the trip as both have knowledge of the routes and know each other's strengths and weaknesses.
10. *Tola* (10 grams) is a unit to measure gold, and musk pods are also measured in the same unit.
11. Marwaris are traders from Rajasthan who run businesses in Arunachal and other parts of northeast India. They frequently travel in and out of the state and engage in trading in animal parts like musk pods.
12. In Anjaw district of Arunachal Pradesh, the Mishmis reported that men from Myanmar give leg-hold metal traps to the local people for trapping otters.
13. There are five clans (Meme, Umpo, Mena, Mischi and Misiwo) of the Idu Mishmis who are exempted from performing this ritual. Members of these clans are believed to be the descendants of the first priest of the Mishmis, Sineru, and therefore, they do not have to perform this ritual, but killing of tigers is still a taboo for them.
14. A healing ritual usually carried out for a day.
15. *Reh* is one of the most important festivals of the Mishmis for propitiation of the supreme creator, the *Nanyi Inyitaya.*

16. A sacred board where the skulls of hunted animals are displayed.
17. T. Rina, 2012, 'Namdapha Tiger Reserve under threat: PCCF bears the Brunt of Poachers' Ire', *The Arunachal Times*, 17 March 2012, available at http://www.arunachaltimes.in/archives/mar12%20 17.html (accessed on 26 February 2017).
18. The Brahma Kumaris is a new religious movement that teaches a form of meditation that focuses on the soul. Their activities promote respect for all faiths and meditation through social service activities. This movement has its headquarters in Rajasthan.
19. AoL (Art of Living) is a volunteer-based spiritual NGO. Its programmes combine the mystical and the modern to help create a life of purpose, joy and confidence. It has its headquarters in Bangalore.
20. The Arunachal Development Board.
21. *Junglee* is a Hindi word derived from *jangal* (wood, forest), applied in a derogatory way to 'tribals' or 'forest-dwelling' people.

References

Aisher, A. 2005. *Through 'Spirits': Cosmology and Landscape Ecology Among the Nyishi Tribe of Upland Arunachal Pradesh, Northeast India,* PhD Thesis (Anthropology), UK: University College, London.

———. 2007. 'Voices of Uncertainty: Spirits, Humans and Forests in Upland Arunachal Pradesh, India', *South Asia: Journal of South Asian Studies* 30(3): 479–98.

Aiyadurai, A. 2007a. *Hunting in a Biodiversity Hotspot: A Survey on Hunting Practices by Indigenous Communities in Arunachal Pradesh, Northeast India.* UK: Rufford Small Grants Foundation; Mysore: Nature Conservation Foundation.

———. 2007b. 'Pheasant Hunting: A Cultural Practice in Arunachal Pradesh, Northeast India', *The International Newsletter of the World Pheasant Association* 79: 6–7.

———. 2009. *Wildlife Hunting in Miju Mishmi, an Indigenous Group in Arunachal Pradesh, Northeast India: A Socio-economic and Cultural Study,* M.Sc. Dissertation (Anthropology), UK: University College London.

———. 2011. 'Wildlife Hunting and Conservation in Northeast India: A Need for an Interdisciplinary Understanding', *International Journal of Galliformes Conservation* 2: 61–73.

———. 2016. '*Tigers are Our Brothers*': *Understanding Human–Nature Relations in the Mishmi Hills, Northeast India,* PhD Thesis (Anthropology), Singapore: National University of Singapore.

Aiyadurai, A., N. J. Singh, and E. J. Milner-Gulland. 2010. 'Wildlife

Hunting by Indigenous Tribes: A Case Study from Arunachal Pradesh, Northeast India', *Oryx*, 44(4): 564–72.

Baruah, S. 2003. 'Nationalising Space: Cosmetic Federalism and the Politics of Development in the Northeast India', *Development and Change* 34(5): 915–39.

Bennett, E. L., and J. G. Robinson. 2000. *Hunting of Wildlife in Tropical Forests: Implications for Biodiversity and Forest Peoples.* Working paper No. 76, Biodiversity Series. Washington, D.C: The World Bank.

Bird-David, N. 1990. 'The Giving Environment: Another Perspective on the Economic System of Gatherer-Hunters', *Current Anthropology* 31(2): 189–96.

Blench, R. 2016. *Hunting Among the Idu, A People of Arunachal Pradesh.* Roing, Arunachal Pradesh. Available at https://www.academia.edu/20315411/Hunting_among_the_Idu_a_people_of_Arunachal_Pradesh (accessed 2 January 2017).

Bourdieu, P. 1986. 'The Forms of Capital'. In J. G. Richardson (ed.), *Handbook of Theory and Research for the Sociology of Education,* pp. 46–58. New York: Greenwood.

Broadbent, N. D., and B. W. Edvingen. 2011. 'Sacred Sites, Settlements and Place-Names: Ancient Saami Landscapes'. In P. Jordan (ed.), *Landscape and Culture in Northern Eurasia,* pp. 315–38. London and New York: Routledge.

Byrne, D. 2010. 'The Enchanted Earth: Numinous Sacred Sites'. In B. Verschuuren, R. Wild, J. McNeely, et al. (eds), *Sacred Natural Sites: Conserving in Nature and Culture,* pp. 53–62. Washington, D.C.: Earthscan.

Chaudhuri, S. 2008. 'Plight of the Igus: Notes on Shamanism Among the Idu Mishmis of Arunachal Pradesh, India', *European Bulletin of Himalayan Research* 32: 84–108.

Das, A. K., and S. P. Shukla. 2007. 'Biodiversity and Indigenous Knowledge System', *Current Science* 92(3): 275–76.

Datta, A. 2007. 'Protecting with People in Namdapha: Threatened Forests, Forgotten People'. In G. Shahabuddin and M. Rangarajan (eds), *Making Conservation Work: Securing Biodiversity in This New Century,* pp. 120–65. New Delhi: Permanent Black.

———. 2015. 'The Hunter'. Available at http://blog.ncf-india.org/2015/04/24/the-hunter/ (accessed 10 November 2016).

Datta, A., M. O. Anand, and R. Naniwadekar. 2008. 'Empty Forests: Large Carnivore and Prey Abundance in Namdapha National Park, North-East India', *Biological Conservation* 141(5): 1429–35.

del Valle, F. B. 2002. 'The National Sanctuary Pampas del Heath: Case Study of a Typical "Paper Park" under Management of an NGO'. In J. Terborgh, C. van Schaik, L. Davenport, and M. Rao (eds), *Making Parks Work: Strategies for Preserving Tropical Nature,* pp. 149–55. Washington: Island Press.

Donovan, D. G. 2004. 'Cultural Underpinnings of the Wildlife Trade in Southeast Asia'. In J. Knight (ed.), *Wildlife in Asia: Cultural Perspectives,* pp. 88–111. London: RoutledgeCurzon.

Dutt, B. 2004. *Biodiversity, Livelihoods and the Law: The Case of Jogi-Nath Snake Charmers of India.* New Delhi: WTI.

Dutt, B., R. Kaleta, and V. Hoshing. 2007. 'The Hunter and the Hunted: Conservation with Marginalized Communities'. In G. Shahabuddin and M. Rangarajan (eds), *Making Conservation Work: Securing Biodiversity in this New Century,* pp. 88–11. Ranikhet: Permanent Black.

Elwin, V. 1964. *The Tribal World of Verrier Elwin: An Autobiography.* London: Oxford University Press.

Gadgil, M., and K. C. Malhotra. 1998. 'The Ecological Significance of Caste'. In R. Guha (ed.), *Social Ecology,* pp. 27–41. Delhi: Oxford University Press.

GoI. 1994. *The Indian Wildlife Protection Act,* 1972 (as amended up to 1993), Second edition, Dehra Dun: Natraj Publishers.

Griffin, P. B., and M. B. Griffin. 2000. 'Agta Hunting and Sustainability of Resource Use in Northeastern Luzon, Philippines'. In J. G. Robinson and E. L. Bennett (eds), *Hunting for Sustainability in Tropical Forests,* pp. 325–35. New York: Columbia University Press.

Hilaluddin, R. Kaul, and D. Ghose. 2005. 'Conservation Implications of Wild Animal Biomass Extractions in Northeast India', *Animal Biodiversity and Conservation* 28(2): 169–79.

Howell, S. 1984. *Society and Cosmos: Chewong of Peninsular Malaysia.* Singapore: Oxford University Press.

Kashyap, S. G. 2016. 'Nagaland Village Erects Monolith for Conserving Migratory Falcons', *The Indian Express,* 21 January 2016. Available at http://indianexpress.com/article/india/india-news-india/nagaland-village-erects-monolith-for-conserving-migratory-falcons/ (accessed 8 January 2017).

Knight, J. 2004. *Wildlife in Asia: Cultural Perspectives.* London: RoutledgeCurzon.

Kumar, A., and B. Riba. 2015. 'Assessment of Effectiveness of Conservation Action for Hornbill Species in Arunachal Pradesh, India: The Great Indian Hornbill (*Buceros bicornis*)', *International Journal of Conservation Science* 6(1): 125–34.

Milner-Gulland, E. J., E. L. Bennett, and the SCB 2002 Annual Meeting Wild Meat Group. 2003. 'Wild Meat: The Bigger Picture', *Trends in Ecology and Evolution* 18(7): 351–57.

Morris, B. 1998. *The Power of Animals: An Ethnography.* Oxford: BERG.

———. 2000. *Animals and Ancestors: An Ethnography.* Oxford: BERG.

———. 2004. *The History and Conservation of Mammals in Malawi.* Malawi: Kachere Series.

Myers, N., R. A. Mittermeier, C. G. Mittermeier, et al. 2000. 'Biodiversity Hotspots for Conservation Priorities', *Nature* 403: 853–58.

PTI. 2013. 'Naga Villages Pledge to Save Migratory Amur Falcons', *Business Standard*, 28 August 2013. Available at http://www.business-standard.com/article/pti-stories/naga-villages-pledge-to-save-migratory-amur-falcons-113082800786_1.html (accessed 6 February 2017).

Rabinowitz, A. 2001. *Beyond the Last Village: A Journey of Discovery in Asia's Forbidden Wilderness*. Washington: Island Press.

Rina, T. 2012. 'Namdapha Tiger Reserve under threat: PCCF bears the Brunt of Poachers' Ire', *The Arunachal Times*, 17 March 2012. Available at http://www.arunachaltimes.in/archives/mar12%2017.html (accessed 26 February 2017).

Sinha, N. 2014. 'Working Together to Face the Crisis', *Mistnet* 15 (2): 22–23.

Tsing, A. 2013. 'More-Than-Human-Sociality'. In K. Hastrup (ed.), *Anthropology and Nature*, pp. 27–42. New York: Routledge.

Velho, N., K. Karanth, and W. F. Laurance. 2012. 'Hunting: A Serious and Understudied Threat in India, a Globally Significant Conservation Region', *Biological Conservation* 148(1): 210–15.

Vidyarthi, L. P. 1963. *The Maler: A Study in Nature-Spirit-Man Complex of a Hill Tribe in Bihar*. Calcutta: Bookland Private Limited.

2

Conservation in Urban Spaces

People–Wildlife Interactions and Management of Delhi's Forests

M. Vikas

Introduction

Encountering an animal in an Indian city has hardly ever been remarkable. Many animals, both domesticated and wild, have long found refuge in the bustling cities of the Indian subcontinent. Langurs and macaques stealing food outside temples, peacocks screaming as the *fajr azaan* calls the devout to morning prayers, emaciated captive elephants lumbering through heavy traffic or cows defiantly provoking traffic jams; the cacophony and drama that plays out on India's streets has long been criticised and romanticised. Since 1991, however, socio-economic changes and other interplaying factors have led to a churning of cityscapes, and a paradigm shift in the social construction of wild spaces. The unprecedented urbanisation and change in land use in peri-urban India is blurring the historically fluid boundaries between human and wild spaces. With cities fast devouring their peripheries, the rural agricultural–forest continuum is fast unravelling because of the loss of both agricultural land and forests and leading to increased human–wildlife interactions. Forest officials in cities, who had hitherto managed environmental concerns like waste disposal and water quality, are today overseeing rescue operations to tranquilise leopards trapped in a human habitation, while also managing crowds agitated in their wake. This is not to say that the presence of leopards within cityscapes is a new phenomenon. Rather, it underlines the fact that few cities have included wildlife conservation in their management plans or have staff equipped and trained to handle wildlife rescue, and most have abjectly failed in protecting urban wildlife habitat. With the United Nations (UN) predicting the global urban population to increase to 5.26 billion by 2050 (Montgomery 2007), conflicts are bound to increase as people and wildlife will compete for the last remaining natural resources. This chapter discusses the management of forested areas in Delhi and its environs, its roots in British colonialism, and explores its current political ecology. The chapter is divided into the seven following sections: Delhi's Landscape; Natural History; Colonial Continuity in Post-colonial India; Post-Independence Conservation Efforts; Impact of Economic Liberalisation on Forests and Commons; and Conclusion with Key Policy Recommendations.

Delhi's Landscape

Geographic and Cultural Landscape

Delhi is a Union Territory (UT) spread over 1,484 sq. km, stretching between 28° 24' 17" and 28° 53' 00" North latitude and 76° 50' 24" and 77° 20' 37" East longitude (Mishra et al. 2015). The Indian National Capital Region (NCR) seamlessly expands into an urban sprawl comprising several neighbouring, suburban 'satellite towns' such as Gurgaon and Faridabad in Haryana, and NOIDA and Ghaziabad in Uttar Pradesh. While the NCR of Delhi accommodates approximately 16.7 million residents (Chandramouli and Registrar General 2011), the combined population of Delhi and the cities that encircle it is about 26.45 million and projected to explode to 36.06 million in 2030 (UN 2016). Apart from putting a severe strain on fast-depleting natural resources, approximately 66–80 per cent of Delhi's surface area is predicted to be built up to accommodate the growing population (Mohan et al. 2011). Noxious fumes billowing from industries, cars and the machinery of industry have cumulatively contributed towards making Delhi's air the world's most noxious in 2016. Despite hosting millions of people and heavily polluted air and water engulfing it, Delhi is also paradoxically one of the world's greenest cities with 20.59 per cent cumulative forest and tree cover (Forest Survey of India 2017), and is one of the few national capitals that have protected areas (PAs) for wildlife within its municipal borders.

Two main geographical features that influence Delhi's varied topography are the Yamuna River flowing along its eastern border and the extensive Aravalli hills that encircled the historical urban centres. There are four broad categories of land, each with unique flora and fauna: (a) the Aravalli hills (ridge, *kohi* or *pahari*), (b) old alluvium plains primarily irrigated by wells (*bangar*), (c) riverine, sandy land (*khadar*), and (d) low-lying catchment area that the Yamuna frequently floods (*dabar*) (Krishen 2006). With light and fertile soil, vegetation in *bangar* areas would have comprised savannah or monsoon forests of northern India, provided the area had been free of human intervention. While the riverine *khadar* areas had typically riparian forests interspersed with dense woody patches, the flood-prone *dabar* areas supported trees that could survive the saline soil (ibid.). The Ridge's flora primarily comprises thorny, secondary forests (*rakh*) that are typical to arid/semi-arid regions of the world (Maheshwari 1963). A small portion of the Ridge is composed of sandy alluvium that supports a lusher microhabitat similar to Western Rajasthan (Krishen 2006). The geology of the region also affects the local vegetation, and villagers identify three kinds of soil based on its consistency: (a) clayey loam (*dakar*), (b) clayey-sandy loam (*rausli*), and (c) sandy loam (*bhur*).

Delhi Ridge marks the tail-end Mewat branch of the Aravalli mountain range, the oldest mountain range in India (Mishra et al. 2015). The Aravalli hills enter Delhi from the southern region and bifurcate into two branches that encircle Delhi's plains like a sabre made of detached successive hills and eroded, extensively quarried outcrops of ferruginous and lime quartzite rocks, grits and schistose rocks along with granite (Maheshwari 1963; Varsha 2014). Symmetrically folded quartzite rocks dominate the Ridge, most probably formed during the tertiary period by the uplift of the strata due to peneplanation. The parallel system of flat-topped synclinal Ridge system is intervened by anticlinal valleys buried under *bangar* alluvium (Dakshini 1968). Once dominating the landscape, today the Ridge forests exist in Delhi in fragments, with several pieces categorised as reserved forests, protected forests, city forests, and/or biodiversity parks, each governed under different statutes by various governmental bodies. The largest contiguous block is the Southern Ridge, a large portion of which the government notified as Asola Bhatti Wildlife Sanctuary (ABWS) under the Wildlife Protection Act, 1972. The Ridge topography is very uneven, with numerous ravines and projecting ribs of rock interrupted by sandy soil with low humus content and covered under sparse vegetation of primarily grasses and shrubs. Soil in this area is coarse, low in organic matter and nitrogen, and is categorised in the world group of grey/brown soils of desert/semi-desert regions. The extreme climate of Delhi that oscillates between very hot summers and cold winters, and other factors like the sharp diurnal difference in daily temperatures, high saturation deficit and moderately low rainfall averaging 60 cm annually, cumulatively make the region unfavourable for luxuriant vegetation, with largely drought-resistant and xerophytic trees being able to survive. Because of a variation in the substratum along with human interventions like grazing, there is significant variation in the vegetative composition of the Ridge (ibid.).

The micro-distribution of tree species in the Ridge area is typical of the Aravallis in north-western India. Champion and Seth classified the Ridge vegetation as semi-arid, scrub forest, although dry deciduous slope and riverine gallery forests are also present (Champion and Seth 1968). The extensive open, plateau-like areas are dominated by thorny scrub savannah associations that include tree species such as *Acacia leucophloea, Acacia nilotica* and *Ziziphus mauritiana*; shrubs such as *jhand* (*Capparis sepiaria*), *Capparis decidua* and *Balanites roxburghii*, and thorny bush-weeds like *pala/beri* (*Zizyphus nummularia*) and *munj* grass (*Saccharum sara*) (Maheshwari 1963). *Dhok* or *dhau (Anogeissus pendula)* is a deciduous species that dominates the steeper slopes of the undulating valleys in the Delhi Ridge, while seasonal stream beds with higher soil moisture harbour evergreen trees such as *Mitragyna parviflora, Holoptelea integrifolia* and *Ficus glomerata. Dhok* and riverine forests are presently limited to very few sites, although they were possibly more extensive in the past. Since *dhok* grows in limited patches, Maheshwari conjectures that its geographical extent may have reduced due to human exploitation of the tree for wood and livestock-feed (ibid.).

Vegetation in the Ridge also varies as one travels southwards; while *Prosopis spicigera* along with *Acacia modesta* and other species dominate the northern edge of the Ridge, *Rhus mysurensis* and other trees dominate the Southern Ridge (Dakshini 1968). Unlike the subsidiary edaphic type of dry, tropical forests of the Aravallis in Rajasthan (sub-type E-6 *Anogeissus*), Champion (1936) records Delhi with a few pockets of the dry, tropical sub-type of *Anogeissus.* Parker (1920) holds that the natural forest cover of Delhi was primarily a scrubby forest of *Anogeissus* and *Acacia senegal.* Additionally, deciduous trees such as *Sterculia urens* and *Boswellia serrata* grew on the rocky 'cliff edges' of the hill, now largely extinct from the NCR. Today, the most widespread tree in the Ridge that can survive in all types of terrain is *vilayati kikar* or *Prosopis juliflora,* a Mexican native introduced in India in 1877 by British foresters (Kalpavriksh 1991). *Prosopis juliflora* has thrived as an invasive species in the Ridge, and recent studies estimate that it covers 23.43 sq. km of the total 32.71 sq. km ABWS (Kushwaha et al. 2014). *Prosopis juliflora* can dramatically alter forested landscapes within years, by crowding out native scrub forest species, both shrubs and herbs, through a high degree of coppicing. A 2015 study recorded Delhi's present floral diversity to include 377 genera and 551 species belonging to 90 angiospermic families, which includes 76 dicots and 14 monocots (Mishra et al. 2015).

Delhi is geographically part of the larger north Indian socio-cultural ecosystem. Much of north India's topography was once vast expanses of semi-arid land 'in the nature of open steppes' (Chakravarty-Kaul 2015), marked by dry climate and erratic rainfall. Natural conditions, complex social relations and institutions like caste, which mediated human–environmental interactions and land use in the region, also policed material culture and human livelihood. Customary rules delineated how different social groups only exploited a certain portion of the land, leaving the rest for another community to exploit it according to its own economic modes. The water-scarce regions in northern India shaped both the agricultural and pastoral economy in the region; farmers grew grains like *bajra,* while nomadic pastoral groups moved with grazing rights in grasslands and forests (Ludden 2011). There was a relationship between the seasonal movement of shepherds and agricultural cycles, and nomadism did not conflict with settled agriculture. Settled agriculturalists had integrated itinerant pastoralists into the rural economy, and they performed complementary roles for other communities (Bhattacharya 1992).

This historical engagement of people with their landscape heavily altered it in the process, resulting in a mosaic of natural and used patches of open canopy, scrub and denser forests. The transformation of Delhi's geographical landscape is also closely entwined with its ever-changing social tapestry, with the city having embraced several waves of migrants like Jats and Meos (Singh 2006). While mainly agriculturalist Jats tilled the plains, semi-nomadic pastoralist Gujjars herded their livestock in the Ridge. As much as these landscapes had myriad anthropogenic uses, wildlife always roamed free in this land. Historically, Delhi was home to a diverse wildlife population, and *Gazetteer of the Delhi District* mentions thriving wildlife that included large herds of blackbuck antelope (*Antilope cervicapra*), Indian gazelle (*Gazella bennettii,* locally called chinkara), and blue bull antelope (*Boselaphus tragocamelus,* locally called nilgai) (Gazetteer of the Delhi District 1884). Many Tughlaq-era hunting lodges such as Malcha Mahal, Bhuli Bhatiyari ka Mahal and Kushak Mahal attest the rich wildlife that would have merited royal hunts (ibid.). Much of Delhi's wildlife has disappeared since then, and last remaining species like jackals and nilgai are holding on to their habitat by the skin of their teeth (Shahabuddin 2016). In fact, the Zoological Survey of India termed ABWS as an 'artificially created', 'man-made' PA, and quotes the World Wildlife Fund, questioning its wildlife sanctuary status in 1995, given the paucity of wildlife (Zoological Survey of India 2003).

Although ABWS is a relatively small PA, it is part of a much larger, still broadly contiguous block of Aravalli forests spread across Delhi, and Gurgaon and Faridabad districts. Despite fragmentation and severe anthropogenic pressure, the undulating landscape continues to function as a wildlife corridor that extends approximately 200 kilometres from Delhi up to Sariska in Rajasthan. However, as human pressure on Delhi's forested purlieus continues to increase, the existence of wildlife in ABWS and neighbouring districts is gravely imperilled. The Delhi state government cannot manage wildlife in ABWS in isolation, and in the absence of a comprehensive wildlife management policy that integrates the entire Aravalli landscape, ABWS may end up being nothing more than a glorified zoo. In addition to similar topography and vegetation carpeting the terrain, southern Delhi and neighbouring districts of Haryana also share history and culture, making this region a larger socio-cultural block that should be assessed holistically (Singh 2006). To conserve the Ridge across Delhi and Haryana, and not just ABWS, it is also important to gain fluency in the cultural, economic and institutional milieu of the region. Understanding important issues can help policy-makers situate conservation policies within the local context. Understanding the social construction of landscapes will help authorities build edifices upon it that are congruent with their natural and cultural rhythms.

For the purpose of this chapter, references to Delhi are not only restricted to the present geographical areas comprising the National Capital Territory of Delhi, but more broadly also include the neighbouring areas of Haryana. The historical cities of 'Delhi' and its borders have been fluid and shifted multiple times. For example, Anangpur, one of the first forts of 'Delhi', is located in present-day Faridabad district, while a stone inscription found in Mehrauli in south Delhi refers to Dhillika (the name of one of the many former cities of Delhi) as a part of Haryana. Therefore, it would be difficult to restrict our analysis to a modern geographical limit or a specific historical period. The historically blurred lines between Delhi and Haryana's boundaries also speak to the closely entwined fates of their landscapes. This chapter seeks to delineate the historical trajectory of conservation in the region and attempts to identify ways in which government authorities can implement wildlife conservation in an environmentally sustainable and socially just manner.

Natural History

The natural history of Delhi until the establishment of Rajput rule from the seventh to eighth centuries is sketchy. Archaeological evidence from the region includes Palaeolithic artefacts all the way through the centuries to the medieval cities of Delhi (Trivedi 2009). Archaeological evidence provides information about land use by Rajputs in this early period, with remains of numerous embankments or *bunds* that collected and channelled water to fields. The Sultanate rulers extended irrigation facilities from the thirteenth to sixteenth centuries and expanded agriculture in the Indo-Gangetic plains (*doab*) for higher revenue (Fuller and Khallaque 1967). Sultanate rulers laid out numerous orchards across the city's environs, with Firoz Shah Tughlaq reportedly planting 1,200 gardens in Delhi (Page 1937). Although orchards did provide aesthetic relief to medieval rulers, horticulture was not the sole purpose. Orchards were also an important source of revenue. For example, fruits grown in Delhi's orchards gave the exchequer more *tankha* than the entire *doab* (Jauhri 2001). Mughal rule until 1857 CE witnessed the laying down of several walled gardens, including the Shalimar Garden, the Raushanara Garden, and the Qudsiya Bagh on the banks of the Yamuna (Sharma 1964). Even after Shahjahanabad became the Mughal capital in mid-seventeenth century CE, large portions of Delhi and its neighbourhood remained wild, and sources like records of Shah Jahan hunting 40 blackbuck in one day in Palam evidence abundant wildlife (Rangarajan 2001).

Early nineteenth-century colonial records show the Ridge as contiguous across Delhi. British presence was initially restricted to the cantonment constructed atop the Northern Ridge (NR); in fact, health concerns of the British led to a special medical committee in 1863 declaring that the old cantonment on the Ridge was a better place for European troops (Gupta 1981). The most important ecological impact of British rule in Delhi was not fragmentation but concerted afforestation in the Ridge from 1883 onwards (Mann and Sehrawat 2009). Delhi's tree cover prior to British colonialism was sparse; Frederick Treves describes the southern portion of Delhi as 'a desolate plain covered with the ruin and wreckage of many cities'. He describes the experience of entering the road starting from Delhi Gate as entering a desert, a country of stones, ruined buildings and dust with shabby bushes around: 'As far as the yes can stretch, the scene is the same—dust and crumbling stone … ', he said (Kaul 1985: 10–17). Edward Lear also writes: 'The particularly desolate sphere of Delhi suburbs and neighbourhood or surroundings only comes home to one after much going about the place; the endless heaps of ruin; the countless tombs; the small amount, at least apparently, of population' (ibid.). The 1857 War of Independence would have also wreaked havoc on Delhi's nature and environment. One must understand that these bleak descriptions of the landscape were based on British understanding of nature's phenotypic qualities; this 'bleak' land was biologically rich enough to support the Gujjar pastoral economy as well as plentiful wildlife (Gazetteer of the Delhi District 1884).

Initial afforestation efforts began near British settlements in north Delhi to reduce local temperature and beautify the bare surroundings. Plantation of new saplings started the long and continuing process of exclusion of local villagers from their lands, especially by restricting grazing rights. After the British selected Delhi as the Imperial Capital in the early twentieth century, afforestation efforts expanded in what is present-day Kamala Nehru Ridge and Central Ridge Reserve Forest (CRRF). In addition to planting native species of flora, British officers encouraged the plantation of foreign species like *Prosopis juliflora* because of their drought resistant properties and visual similarity to the indigenous *babul*. The British hoped to create a majestic 'sea of foliage' in tune with the worldwide pattern of increasing greenery in urban areas: 'Our real difficulty in new Delhi is not going to be to hide ugly things, but to prevent what we want to be seen from being hidden; while keeping the city green and tidy' (Hasan and Patel 2014).[1] Many scholars believe that colonial officers implemented this urban forestry project for the benefit of the Europeans at the expense of local villagers (Mann and Sehrawat 2009). The reconstruction of the ecology of the Ridge was an intrinsic part of making imperial Delhi; while Lutyens laid out the city of the empire, foresters selected trees to afforest the barren slopes of the Ridge. Given that the British had no intentions of expanding agriculture in the Ridge, one can reasonably conclude that the long-drawn afforestation programme was not an economic project, but rather, an aesthetic one. Apart from Delhi's forests, colonial rule (and the 1947 Partition) also fundamentally altered Delhi's rural character. In 1901, 52.76 per cent of its 0.4 million citizens lived in urban areas, which increased to 82.40 per cent of 1.74 million people in 1951 (Mohan et al. 2011).

British colonialists had strong economic interests in land, with revenues extracted from the production and sale of agricultural and forest goods. With greater value attributed to farms and timber forests, the British deemed other kinds of lands such as value-producing pastures and village commons as wasteland (Wells 2007; Whitehead 2012). In fact, Malthus opined that the establishment of private property over wastelands was critical to increase agricultural production in India. British modification of the public space and its boundaries began the process of changing institutions that governed common property resources (CPRs). In pre-British times, villages across north India used village fallows in an 'open field system' that had specific economic uses throughout the year. The village commons in Delhi, for example, was the *shamlat/shamilat-deh.* CPRs primarily acted as a public space used by herders, with small agrarian settlements providing much needed food and water to the pastoralists (Chakravarty-Kaul 2015). Networks of CPRs were spread over large geographical areas, with their governance necessitating property rights of communities over these lands. British revenue records list detailed customary rules that set out the manner of use of land for cultivation and grazing, as well as economical use of water. These rules helped the local village communities in sustainably exploiting natural resources and spreading the economic risks to both agriculture and grazing due to climatic fluctuations. Since all communities were inter-dependent and living off the same land, CPRs symbolised their joint stake in the continuation of the institutions that governed them.

British intervention not only led to the extinguishment of many land rights, but also transformed land use and through it, transformed the ecology of the region. The carefully balanced pastoral and agricultural land uses changed towards more agriculture as canals expanded, making more land cultivable. The increase of cattle stock also put pressure on the grazing lands, which fell short of the required biomass to support cattle numbers, leading to overgrazing of commons (Chakravarty-Kaul 2015). Breakdown of traditional institutions, loss of knowledge and increase in both human and livestock population ultimately caused degradation of CPRs and created more wasteland. The British also took initiatives to sedentarise pastoral communities since the State was interested in expanding agricultural land. Since the experience of colonialism by Indians was mediated by caste and tribe identities (Guha 1999), pastoral communities identified with 'unproductive' CPRs came to be viewed pejoratively (Gidwani 1992). In addition, provisions of the Criminal Tribes Act (CTA), 1871 criminalised many itinerant communities, nomadic and semi-nomadic communities, as well as settled agriculturists.[2] With a change in the economy, traditional ties also began to fray, with communities responding variously by either restricting access to commons by way of grazing fees or disbarment of outsiders, or by privatisation of the commons itself. British colonialists were not responsible for the final undoing of Delhi's landscape; it was and continues to be carried out by Indians.

Colonial Continuity in Post-Colonial India

Delhi's post-Independence environmental saga has recurring colonial characteristics like marginalisation of CPRs and pastoralism, disdain for arid ecosystems and perpetuation of unequal habitation patterns. While the British commenced the process of modifying notions of public space, the process of changing the institutional governance of CPRs accelerated after Independence as the State took a more direct interest in issues of land ownership and governance. With the enactment of the Land Reforms Act, 1947, the Panchayati Act, 1954, and the Village Common Lands Act, 1961, scholars believe that the State effectively dismantled the CPR systems by making CPRs more open access (*gaon sabha* land), ultimately leading to its environmental degradation. CPRs were also privatised and redistributed among landless Dalit households, in some instances to correct historical injustices committed against these communities. Earlier, dominant castes had accommodated Dalits as tenants-at-will into village communities on conditions like providing complementary services. Apart from getting residential rights, Dalits also had access to CPRs. The tenants-at-will cultivated a substantial portion of the agricultural land (32.5 per cent from 1906–10), making them an important component of the village economy. Some scholars believe that although the goal was noble, land reforms were problematic in praxis since the government created private property by usurpation and redistribution of CPRs (Chakravarty-Kaul 2015).

The biggest catastrophe for Delhi's CPRs and agricultural land, however, was the influx of 5,00,000 refugees after Partition, which led to unplanned constructions across the city, including over the Ridge. The Delhi Development Authority (DDA) was established in 1957 to regulate the establishment of settlements and provide affordable housing to citizens (Sheikh and Mandelkern 2014), and plan the city's orderly expansion.[3] The DDA actively purchased and acquired agricultural and Ridge land, increasing the built-up area in Delhi from about 540 sq. km in 1997 to almost 792 sq. km in 2008. Approximately 258 sq. km of agricultural land, 'wasteland' and water bodies correspondingly decreased during that time (Mohan et al. 2011). However, the devastation of CPRs and agricultural land did not achieve the goal of providing affordable housing to people in Delhi. Despite the DDA controlling 365.5 sq. km in Delhi, it failed to provide planned housing to people. Between 1951 and 2001, while population grew from 1.7 million to 13.8 million, approximately 50 per cent of this population was residing in informal settlements (Véron 2006).

Apart from converting forested land into residential areas, the Delhi administration also continued the parallel process of conservation of portions of the Ridge. Initially, the DDA managed all fragments of the Ridge. However, over the years, the Delhi administration distributed this role among numerous government bodies; resultantly, the trajectories of governance of the fragmented Ridge also went in different directions. The primary focus of the government in the early years was on the NR and CRRF. In 1968, the DDA took charge of the NR (Varsha 2014), and plantations and fencing continued at both NR and CRRF over the next few years. In 1962, the Central Public Works Department (CPWD) took over the maintenance of the Southern Ridge, which began creating parks and gardens in earnest. The first Master Plan for Delhi proposed designating and protecting the Ridge as a forest or regional park, broadly on the lines of New York's Central Park. The plan recognised the ecological value of the Ridge and recommended that the hills should be retained in an 'undisturbed' state, while contradictorily also recommended regular weeding and clearing of undergrowth. In addition, it did not spell out clear guidelines or steps about how to carry out this plan (Kalpavriksh 1991). These confusions and contradictions at the policy planning stage led to the Delhi administration not notifying or implementing formal policies to treat and manage the Ridge as a wildlife habitat, ultimately leading to its slow destruction.

In the following decades, construction activities in the Ridge continued unabated. Several unauthorised colonies came up in the Ridge, built by not only poor migrants, but also Delhi's economic elite (Sheikh and Banda 2014). Various government bodies allowed the construction of buildings and roads throughout the Ridge. The horticulture departments of bodies like the DDA and the New Delhi Municipal Council (NDMC) focused on clearing undergrowth and setting up parks over the Ridge instead of managing the space for wildlife (Kalpavriksh 1991). The continued deterioration of the Ridge ultimately precipitated citizen action, with a series of demonstrations in 1979 led by people, prominent citizens and non-governmental organisations (NGOs) like Kalpavriksh, demanding protection of the Ridge. Succumbing to public pressure, in 1980, the Lieutenant Governor of Delhi declared 25 sites in the NR, CRRF and South–Central Ridge as Protected Forests under the Indian Forest Act, 1927 (Varsha 2014). By the time the Delhi administration announced these progressive steps, much of the Ridge had already been devastated. In 1982, a 'secret' note from the Town and Country Planning Organisation found that 40 per cent of the Ridge had been destroyed; much of it due to the government diverting forested land for non-forest uses (Kalpavriksh 1991).

While developments so far had primarily impacted the NR, CRRF and nearby communities, conservation efforts beginning in the 1980s had a profound impact on the hitherto ignored Southern Ridge. Since the British colonial government had declared the NR and CRRF as Reserved Forests in the early twentieth century, those areas had been fenced off and movement of the local villagers had been restricted for several decades. The Southern Ridge, however, remained under the collective ownership of villages as *gaon sabha* land and residents used the land primarily for grazing livestock and quarrying. Mining commenced in the Bhatti region of the Southern Ridge in 1969, attracting labourers from the Odh community, an impoverished, landless Hindu community that had migrated from Sindh after Partition. Odh families settled in three colonies: Sanjay Colony, Indira Nagar and Balbir Nagar around the Bhatti mines. After noted environmentalist M. C. Mehta filed a petition against deteriorating environmental conditions in the city, the courts discontinued mining activity in the Southern Ridge. Because of the extensive quarrying that had carried on in the area for almost two decades, scientists recommended banning extractive practices to begin the process of ecological recovery of the area. Therefore, in 1989, the Delhi administration decided to also notify some portions of the forests as a wildlife sanctuary as per the Wildlife Protection Act, 1972, and 1,882.80 hectares of *gaon sabha* land (out of 6,200 hectares of the Southern Ridge) was notified as Asola Wildlife Sanctuary. Subsequently, the government also alienated 866.5 hectares of contiguous *gaon sabha* land and merged it with Asola Sanctuary in 1991 to create ABWS.

During these years, both the Delhi Administration and an active judiciary managed the governance of the Ridge. For example, the Supreme Court had directed the Delhi government to constitute a Ridge Management Board (Sinha 2014), a body that continues to play a role in protecting the environment in Delhi. One of the key problems in managing the Ridge was its categorisation under different laws; some parts were reserved forests as per the Indian Forests Act, other portions were wildlife sanctuary, whereas the Revenue Department recorded portions as non-cultivable land or *gair mumkin pahar.* These multiple and unclear records had led to confusion regarding ownership and rights over the Ridge. To clarify this position, the Supreme Court identified 7,777 hectares of the 'morphological' Ridge as Reserved Forests as[4] per the Geological Survey of India's study. In 1996, 4,207 hectares of the remaining 'surplus' 'uncultivated' *gaon sabha* land in the Southern Ridge was also transferred to the Forest Department.

Despite almost two decades having passed since ABWS was notified, the government lacks a clear boundary or an authoritative map, and villagers have filed cases against the usurpation of *gaon sabha* land by the government. The original intention was to create ABWS as a deer or tiger safari, with the possible introduction of carnivores. A few blackbuck were introduced in ABWS in zoo-like enclosures (Zoological Survey of India 2003). Species like spotted deer or *Axis axis* (locally called chital) have also been introduced, although the species does not have a recorded presence in the Ridge. Since October 2000, the floral rehabilitation of the area has been carried out by an Eco Task Force, which comprises army personnel. Sustained conservation activities in ABWS over the years have produced positive results. In the last few years, leopards have been increasingly sighted in and around ABWS, although one of them was found impaled on the spiked high wall of one of the farm houses that have cropped up on the Ridge.

Densification of Tree Cover

The ecological success of reviving the Ridge landscape has primarily been possible through extensive afforestation of largely ecologically inappropriate species. It is widely debated that the current forest governance standards applied in semi-arid contexts in India have originated from the 'scientific forestry' paradigm imported from Europe (Gautier et al. 2015), which emphasises timber. Such afforestation activities are not limited to Delhi alone, and tree plantations have been carried out in 8.86 million hectares of wasteland from 1985–90 in the country, aimed towards fulfilling India's goal of 33 per cent forest cover under the National Forest Policy, 1988 (Balooni and Singh 2007), even using species such as eucalyptus and poplar. The mixed arboricultural practices of the DDA and Forest Department have led to a perpetuation of British practices, and forests continue to be defined as medium or dense canopy tree cover. In some cases, where monoculture or exotic species were planted on degraded land, afforestation caused more harm than good (Hunter et al. 1998). The government extensively planted salt-resistant trees like *Prosopis juliflora* to combat the problem of increasing soil salinity and expansion of wasteland (Bhojvaid and Timmer 1998). Although *Prosopis juliflora* plantations broadly survived, leading to an increase in carbon and nitrogen content and reduction in salinity in the soil, it led to a gradual reduction in the biomass of understorey vegetation (ibid.). However, there are other studies that contradict these claims and show that *Prosopis juliflora* facilitates undergrowth, and that grazing and logging have proven to be more ecologically harmful than *Prosopis juliflora* (Naudiyal et al. 2017).

Densification of vegetation has also been pursued because of benefits like reduction of surface temperature (Kant et al. 2009), and planting green belts helped in reducing air pollution, noise pollution and particulate matter (Tyagi et al. 2006).[5] These plantation activities primarily aim to serve urban populations. The Delhi Forest Department has aimed to derive legitimacy for densification of trees by claiming to replant 'original' indigenous trees and remove the ecologically harmful, invasive *Prosopis juliflora*. However, instead of attempting an ecological restoration of the native semi-arid scrub forest, most of the plantation drives aim to increase the canopy density of trees. The original vegetation of the Delhi Ridge especially was widely spaced without a continuous canopy (Dakshini 1968), but this natural characteristic of Delhi's landscape is being deliberately undone in the name of ecological authenticity. Similar plantation drives have been undertaken in Aravalli Biodiversity Park (ABP) in southern Delhi, which is under the jurisdiction of the DDA, and plantations are ongoing in ABWS and adjoining forests in the Tilpat area. The results of afforestation drives are visible in Forest Survey results. After Delhi's forest-tree cover had reduced from 4.78 per cent in 1987 to 1.6 per cent in 1992 (Mishra et al. 1994), government records claim that the number has increased from 1.48 per cent in 1993 to 20.22 per cent in 2015. Of this, 5.27 sq. km is very dense forest, while approximately 44.39 sq. km is moderately dense (Forest Survey of India 2013). These numbers, however, appear to be suspect since much of the Ridge was already covered by dense *Prosopis juliflora* before the native species plantation drive was undertaken. It is possible that the increase in tree cover is simply the result of higher resolution satellite imagery that is now available. Further, if looked at in isolation, these numbers are also misleading. Although the total forest cover has increased, studies that used remote-sensing to measure deforestation observed that stable vegetation cover decreased across the Delhi region from 1986–2010 (Paul and Nagendra 2015). Some percentage of the increased green cover can be attributed to the extensive afforestation on *gaon sabha* lands that have taken place in the peripheral areas of Delhi.

In addition to afforestation in the Ridge, tree plantation efforts in exclusively urban areas have also been patchy in their implementation and have actually led to a reduction in tree diversity and composition in Delhi's parks and neighbourhoods. A recent study compared the composition of trees in Lutyen's Delhi and Dwarka, a neighbourhood in north-western Delhi (Bhalla and Bhattacharya 2015). The study enumerated 4,380 trees in the Lodhi Gardens Park in Lutyens' Delhi, while the researchers recorded only 451 trees in all the parks sampled in Dwarka. In addition, species diversity has significantly declined, with 125 species in New Delhi and only 26 in Dwarka (ibid.).

Given the natural scrub forest vegetation of the Delhi Aravallis, the visible greenery of dense forests is being created as a mental category for the populace, although it never actually existed. This is a practice that has been carried out in neighbouring Rajasthan as well (Robbins 1998). Culture is an adaptive trait and the plantation of new trees can constitute transporting landscapes across spatial boundaries. Planting trees on *gaon sabha* land to create city forests has helped to increase forest cover statistically, but it does not provide ecological benefits. The Forest Department's selection of species for plantation is often determined by survival rate and speed of growth. *Acacia tortilis* or *Israeli babool* has been planted because of its sturdiness and non-preference for foraging by livestock, but it does not encourage any biodiversity to prosper. *Prosopis juliflora* makes the soil infertile (Sharma and Pal 2001) and strongly outcompetes native tree species by crowding out their seedlings and saplings. Although *Prosopis juliflora* does provide woody biomass (El-Keblawy and Al-Rawai 2005; Kushwaha et al. 2014), it is not favoured by the local people, nor are its leaves eaten by livestock. For these reasons, it is ecologically better to plant local species selected according to the microclimate of the area. For example, Garhi Mandi is established on old alluvium where planting *jhand* trees helps to bind the sandy soil, while also providing non-timber forest products and acting as an important source of fodder. People in these arid areas use trees not only for their wood, but they also use their fruit, sap, bark and leaves, which are integrated into their local economies (Gold and Gujar 2002). For example, Gazetteer of the Delhi District reports that *pala* bush-weed was not only an important source of fodder for livestock, but also had commercial value and it was harvested twice a year for sale (Gazetteer of the Delhi District 1884).

Complete closure of village pasture land in erstwhile scrub savannah landscapes had a profound impact on Gujjars. Urbanisation was enforced in villages without providing any alternate means of employment, sources for fodder or economic support to help transition from pastoralism. The rural character of some villages altered radically after the DDA acquired vast tracts of agricultural land for development. State-induced loss of agricultural land later became a reason to justify exclusion of people from forests since they had lost their 'rural' character (Jain 1994). Even in 1994, more than 100,000 people and 5,000 cattle lived in and around ABWS in villages like Bhatti and Asola (ibid.). The livestock population in Delhi has risen by 17 per cent since 1992 (373,000 as per Livestock Census 2003, Government of India [GoI] 2003) however pastures and village commons have become more inaccessible, pushing cows on choked Delhi roads to feed on garbage instead. The Delhi government has recently mandated that dairy farms should be moved out because of their incompatibility with urbanisation. As villages like Asola and Bhatti are being urbanised simply because of proximity to the capital, there is simultaneous pressure on residents to 'urbanise' their means of subsistence as well.

Impact of Economic Liberalisation on Forests and Commons

After accelerated urbanisation following Partition and the settlement of refugees from Pakistan, the next rush towards expansion into the Ridge commenced after liberalisation of the Indian economy from the early 1990s. Most of the Ridge in Gurgaon and Faridabad was originally owned by proprietary families as CPRs, and after land reforms, the land was collectively managed by panchayats. In the 1980s, *gaon sabha* Ridge land was privatised and distributed among all landowning villagers. Land privatisation in Haryana was contemporaneous with planning Delhi's ongoing expansion towards Gurgaon and Faridabad (Jolly 2010). The lack of space and population pressure in Delhi was ultimately transferred to the peripheral areas, leading to fragmentation and deforestation (Paul and Nagendra 2015). Population, however, is not the only driver of deforestation here, but also the changing consumption patterns of India's emerging urban elite that covet space as a mode of surplus generation. Attributing value to space (and as a corollary, assigning value to the use of the space) gives space a commercial value that can be traded (Lefebvre 1991). Thus, hitherto seen as 'wasteland', CPRs of Gurgaon and Faridabad became valuable commodities. Many well-heeled families and corporate groups bought up these lands in anticipation of growing real estate values and craving for a rustic environment created by holiday homes or farm houses.

Real estate interests quickly sensed profits in the offing and bought large swathes of land, including traditional sacred groves like Mangar Bani. Most of these lands, however, were protected under the provisions of the Punjab Land Preservation Act, 1900. In 2012, the then Director General, Consolidation of Holdings, Haryana denotified the consolidation and sale of thousands of acres of Ridge land sold for real estate development. In this case, the provisions of the East Punjab Holdings (Consolidation and Prevention of Fragmentation) Act, 1948 were used to acquire large parcels of land. The aim of the Consolidation Act was to consolidate landholdings for the betterment of agriculture and to protect commons. However, the provisions of the act were misused for real estate development of non-agricultural land. The Principal Bench of the National Green Tribunal took *suo motu* cognisance of the matter,[6] ultimately leading to orders being passed, prohibiting all commercial activity in this forest area.[7] Documents submitted before the courts include details of several other similar land sales and allude to connivance between the Revenue Department and real estate moguls.[8]

Even as village commons were being siphoned off, recreational conservation took firm hold in Gurgaon. Once an open expanse of hilly rocks and agricultural fields, new Gurgaon is now being rebuilt in fundamental ways for its urban elite. The historical open spaces of Gurgaon's landscape are today being populated with the verdant vegetation of new settlements. Often, the use of a space can change or be appropriated, with new ontological construction and importance given to aesthetics (Lynch 1960). In such situations where land contains historical significance for local communities, surrogate discourses can redefine land's meaning by considering landscapes to be empty spaces. Once embodying the 'rural–urban continuum', Gurgaon today is the petri dish where urban experiments of appropriation are changing the landscape, character and governance of the city (Gururani and Kose 2015). Since India is predominantly rural and agricultural, expansion of peri-urban areas into rural India is as much about agrarian transformation as it is about urbanism. Gurgaon's transformation can be gauged by the fact that population in the city has gone up almost 20 times, and agricultural area has decreased from 80 to 26 per cent, while built-up area has increased from 9 to 66 per cent (Gururani 2018).

To compensate for these building activities, Gurgaon authorities have started creating green spaces with large scale 'forest restoration' in the newly created Gurgaon ABP. ABP was formerly *chargah* (grazing land), but today, it is primarily used for jogging and cycling. ABP epitomises notions of aesthetics and neoliberal aspirations for urban life and conservation finding expression and precedence over rural livelihoods. ABP is currently being managed by a local NGO, and its members have planted diverse plants and trees in a sensitive manner, maintaining native variations in species, grasslands and forest associations. ABP's ability to maintain the landscape in a protected environment has certain ecological benefits, as intensive human use tends to homogenise it. In addition, some conservationists say that had the land not been converted into a park, it would have been parcelled and sold off like the rest of the Ridge. The maintenance of ABP as a constructed forest is a better option than watching the rise of a concrete jungle.

The slow land use change of the Aravalli CPRs, whether because of conservation or construction, has harmed both the environment as well as vulnerable communities. Many landless Dalit families had access to CPRs under customary and local laws in the past. However, Dalit families are finding it increasingly difficult to manage their livestock and agriculture because of increasing pressure on natural resources and shrinking CPRs. While pastoral land was turned to be used for other purposes like construction or agriculture, size of livestock holdings has steadily declined. With both agricultural and grazing lands declining in size, along with access to water sources, it has impacted households holding small parcels of land and landless households more acutely. These developments have resulted in their shift towards other employments (Vij and Narain 2016). Threats to forests have also caused some consternation because of religious reasons. Gujjars consider parts of the Ridge as sacred groves, and one such grove called Mangar Bani became part of an agitation against the slow usurpation of forests. Contiguous with ABWS, the existence of these stretches of Aravallis is crucial for the ecological health of Delhi's forests as well. Although a 15.5-kilometre eco-sensitive zone was notified in the Southern Ridge, with 1-kilometre radius around ABWS where commercial and industrial activities are prohibited,[9] the directive has not been implemented and is applicable to contiguous forests in Gurgaon. Pursuant to prolonged local agitation by both villagers and conservationists, and petitions being filed in courts, the Haryana government recently declared 274 hectares of Mangar Bani and its buffer area of 512.7 hectares as a no construction zone.[10] However, threats to the forests are still looming large, with the corporate–political nexus keen to exploit these natural resources. Although judicial interventions have granted protection to small patches of forests for the time-being, these will not ensure the ecological integrity of the Ridge. Away from contentious areas like Mangar, forest land is constantly being chipped away by corporations, threatening wildlife in the region. In fact, places like Manesar have been witnessing several instances of poaching of leopards and other wild animals.

Conclusion

Ancient Delhi has survived millennia of human and climatic vicissitudes, but the environmental tragedy that engulfs it threatens the future survival of the city. Any measures to protect Delhi, even if only for its urban elite, needs to begin with the protection of its air, water and land. Although the Ridge has received statutory protection by various authorities including the Ministry of Environment, Forest and Climate Change (MoEFCC), DDA, and both the Delhi High Court and Supreme Court, degradation of the Ridge has continued unabated. A key reason for this ensuing tragedy is the multiplicity of government bodies that manage the Ridge. This federation of warring tribes includes the DDA, the NDMC, the CPWD and the Land and Development Office. Although not holding Ridge land, bodies like the TCPO and Delhi Urban Arts Commission also have a role in the management decisions (Kalpavriksh 1991). The Forest Department has been reduced to being just another organisation managing the Ridge; the Department controls a mere 2 hectares of the NR as a nursery.

There are departmental differences in management practices as well, which stems for their distinct envisioning of what an urban forest should be. Although Master Plan 2021 envisages the Southern Ridge to be a contiguous block connecting ABWS with Sanjay Van (South–Central Ridge), the DDA has earmarked the Southern Ridge for various non-forest 'green' purposes like golf courses, race courses and gardens. Since the DDA controls 5,050 hectares of 'green areas' and most of the City Forests, it is an important actor. However, their maintenance of forests like Sanjay Van has been criticised by NGOs (Chauhan n.d.). Given the current confusion, some scholars suggest that the only solution is for the Delhi Forest Department to govern the entire Ridge. This argument does have merit since it would reduce administrative confusion and contradictory actions, as well as transfer Ridge land to a department specifically set up for managing forests.

It is unfortunate that while the Ridge has been envisioned as a recreational landscape, it has been forgotten as a lived space that was integral to the economy of Gujjar and other households in the Aravalli area. For generations, the forest provided fuelwood and fodder for village communities. With increased urbanisation, the number of households that use fuelwood for cooking has sharply declined in many parts of the country, in no small measure because of the use of liquid petroleum gas (LPG) cylinders. The number of rural households in Delhi that used LPG increased from 206 per 1,000 households in 1993–94 to 877 in 2004–05 (DeFries and Pandey 2010). However, this did not mean that households in rural Delhi stopped using firewood. The study showed that the number of urban households in Delhi that used firewood and chips decreased from 21 per 1,000 households in 1993–94 to 12 in 2004–05. However, the number of rural households in Delhi that used firewood increased by minor percentage points from 13 per 1,000 households in 1993–94 to 14 in 2004–05 (ibid.). Ever since the notification of ABWS, the needs of natural resource-dependent people have largely disappeared from the discourse on urban conservation in Delhi. These communities and their needs simply are not urban enough for the Green Delhi of the twenty-first century. Forest management needs to reject the binary understanding of human and natural systems as mutually exclusive and distinct.

There is also an urgent need to look beyond target-oriented and carelessly planned tree planting and afforestation. The native scrub savannahs and dry forests of Delhi's landscape performed numerous socio-ecological functions and supported wildlife endemic to semi-arid areas. Delhi's landscape needs to be representative of these historical and ecological facts, as well as consider the forest's role in the economic functioning of marginalised communities. For Delhi's reimagined wilds to be justly recreated, the distinct material and cultural meanings of the Ridge for its diverse populace need to be recognised. Instead of only planting exotic trees, Delhi and Haryana need to jointly protect and revive water catchment areas in the Ridge,[11] particularly through reviving the locally endangered native scrub savannah associations. Given the favourable hydrology of south Delhi, it is important that the government conserve the entire Aravalli stretch, constituted of agriculture–forest–suburban mosaics, from ABWS to Gurgaon city, as a critical area for groundwater recharge (Rao et al. 1996). Synergies among different ecosystem functions have sustained civilisation in this semi-arid diverse landscape for eons, and their recognition would serve well the goal of sustainable and equitable forest conservation in this region.

Notes

1. Letter with note of Captain George S. C. Swinton from Sir James H. Du Boulay to H. Wheeler, Secretary, Government of India, Home Department, on the layout of imperial Delhi, 10 January 1913.
2. While the Criminal Tribes Act was repealed after Independence, these communities were re-stigmatised through the enactment of the Habitual Offenders Act, 1952. Although many international bodies have demanded the repeal of the Act, it has not been achieved yet.
3. The first Master Plan for Delhi was approved in 1962.
4. Notification No. F.10(42)-1/PA/DCF/93/2012-17(I).
5. Govt. of NCT of Delhi, 'Delhi and Its Forests, Department of Forests and Wildlife', pp. 21, available at http://forest.delhigovt.nic.in/1-21.pdf (accessed on 11 October 2016).
6. Application No. 4 of 2013 (*suo moto*), Tribunal on its own Motion v. State of Haryana and Others.
7. Order dated 12 April 2013.
8. Order in the matter of land in Gujjar village, Manesar Tehsil.
9. MoEF, 2016, Draft Notification S.O.2114(E), 15 June 2016.
10. Govt. of Haryana, Town and Country Planning Department, 2016, Notification No. CCP(NCR)/FBD-/CA/2016/1636, 13 June 2016.
11. MoEF, Draft Notification S.O.2114(E).

References

Application No. 4. 2013. (*suo moto*) Tribunal on its own Motion v. State of Haryana and Others.

Balooni, Kulbhushan, and Katar Singh. 2007. 'Prospects and Problems of Afforestation of Wastelands in India: A Synthesis of Macro- and Micro- Perspectives', *Geoforum* 38(6): 1276–89.

Bhalla, Priya, and Prodyut Bhattacharya. 2015. 'Urban Biodiversity and Green Spaces in Delhi: A Case Study of New Settlement and Lutyen's Delhi', *Journal of Human Ecology* 52(1–2): 83–96.

Bhattacharya, Neeladri. 1992. 'Colonial State and Agrarian Society'. In Burton Stein (ed.), *The Making of Agrarian Policy in British India, 1770–1900.* New Delhi: Oxford University Press.

Bhojvaid, P. P., and V. R. Timmer. 1998. 'Soil Dynamics in an Age Sequence of Prosopis Juliflora Planted for Sodic Soil Restoration in India', *Forest Ecology and Management* 106(2): 181–93.

Chakravarty-Kaul, Minoti. 2015. 'Village Communities and "Publicness" in Northern India: Self-Governance of Common Property Resources and the Environment, 1803–2008'. In Haruka Yanagisawa (ed.), *Community, Commons and Natural Resource Management in Asia*, p. 272. Singapore: NUS Press.

Champion, H. G. 1936. 'A Preliminary Survey of Forest Types of India and Burma', Indian Forest Records (New Series), *Silviculture* 1(1): 286. New Delhi: Government of India Press.

Champion, H. G., and S. K. Seth. 1968. 'A Revised Survey of the Forest Types of India'. New Delhi: Government of India Press.

Chandramouli, C., and Registrar General. 2011. 'Census of India 2011', Provisional Population Totals. New Delhi: Government of India.

Chauhan, Aastha. n.d. 'Status of Sanjay Van. A Preliminary Report, (November 2010–May 2011)'. New Delhi: Srishti.

Dakshini, K. M. M. 1968. 'Vegetation of Delhi State: Mehrauli Region. I. Analysis of Pattern', *Proceedings of the National Institute of Sciences in India* 34 (B) (3. FA – 32-2322): 127–33.

DeFries, R., and D. Pandey. 2010. 'Urbanization, the Energy Ladder and Forest Transitions in India's Emerging Economy', *Land Use Policy* 27(2): 130–38.

El-Keblawy, A., and A. Al-Rawai. 2005. 'Effects of Salinity, Temperature and Light on Germination of Invasive *Prosopis juliflora* (Sw.) DC', *Journal of Arid Environments* 61(4): 555–65.

Forest Survey of India. 2013. 'India: State of Forest Report 2013'. New Delhi: Ministry of Environment, Forest and Climate Change.

———. 2017. 'India: State of Forest Report 2017'. New Delhi: Ministry of Environment, Forest and Climate Change.

Fuller, A. R., and A. Khallaque. 1967. *The Reign of Alauddin Khilji Translated from Zia-Ud-Din Barani's Tarikh-I-Firuz Shahi.* Calcutta: Pilgrim Publishers.

Gautier, Denis, Claude Garcia, S. Negi, et al. 2015. 'The Limits and Failures of Existing Forest Governance Standards in Semi-Arid Contexts', *International Forestry Review* 17(S2): 114–26.

Gidwani, Vinay K. 1992. 'Waste and the Permanent Settlement in Bengal', *Economic and Political Weekly* XXIX(04): PE39–PE46.

GoI, Home Department. 1913. Letter with note of Captain George S. C. Swinton from Sir James H. Du Boulay to H. Wheeler, Secretary, 10 January 1913.

GoI. 2003. '17th Indian Livestock Census'. Chennai: Department of Animal Husbandry, Dairying and Fisheries.

Gold, Ann G., and Bhoju R. Gujar. 2002. *In the Time of Trees and Sorrows. Nature, Power and Memory in Rajasthan*, p. 403. London: Duke University Press.

Govt. of Haryana, Town and Country Planning Department. 2016. Notification No. CCP(NCR)/FBD-/CA/2016/1636, 13 June 2016.

Govt. of NCT of Delhi. 2016. 'Delhi and Its Forests, Department of Forests and Wildlife', p. 21. Available at http://forest.delhigovt.nic.in/1-21.pdf (accessed 11 October 2016).

Guha, Sumit. 1999. *Environment and Ethnicity in India, 1200–1991.* Cambridge: Cambridge University Press.

Gupta, Narayani. 1981. *Delhi Between Two Empires 1803-1931: Society, Government and Urban Growth.* Calcutta and Delhi: Eastend Printers and Oxford University Press.

Gururani, Shubhra. 2018. 'When Land Becomes Gold: Political Ecology of the Commons in an Urbanizing Frontier'. In Sharlene Mollett and Thembela Kepe (eds), *Land Rights, Biodiversity Conservation and Justice: Rethinking Parks and People*, pp. 107–25. Routledge. Taylor and Francis Group.

Gururani, Shubhra, and Burak Kose. 2015. 'Shifting Terrain: Questions of Governance in India's Cities and their Peripheries'. In Pierre Hamel and Roger Keil (eds), *Suburban Governance: A Gobal View.* Toronto: University of Toronto Press.

Hasan, M., and D. Patel. 2014. *From Ghalib's Dilli to Lutyen's New Delhi: A Documentary Record.* New Delhi: Oxford University Press.

Hunter, I. R., M. Hobley, and P. Smale. 1998. 'Afforestation of Degraded Land—Pyrrhic Victory over Economic, Social and Ecological Reality?', *Ecological Engineering* 10(1): 97–106.

Jain, Rimjhim, 'Fenced Out', *Down to Earth,* 15 December, 1994.

Jauhri, R. C. 2001. *Medieval India in Transition: Tarikh-I-Firoz Shahi: A First Hand Account.* New Delhi: Sandeep Prakashan.

Jolly, U. S. 2010. *Challenges for a Mega City: Delhi, a Planned City with Unplanned Growth.* New Delhi: Concept Publishing Company.

Kalpavriksh. 1991. *The Delhi Ridge Forest Decline and Conservation.* New Delhi: Kalpavriksh.

Kant, Yogesh, B. D. Bharath, Javed Mallick, et al. 2009. 'Satellite-based Analysis of the Role of Land Use/Land Cover and Vegetation Density on Surface Temperature Regime of Delhi, India', *Journal of the Indian Society of Remote Sensing* 37(2): 201–14.

Kaul, H. K. 1985. 'Ruins and Landscapes'. In Hari Krishen Kaul (ed.), *Historic Delhi: An Anthology*, pp. 10–17. New Delhi: Oxford University Press.

Krishen, Pradip. 2006. *Trees of Delhi: A Field Guide.* Delhi: Penguin Books India.

Kushwaha, S. P. S., S. Nandy, and Mohini Gupta. 2014. 'Growing Stock and Woody Biomass Assessment in Asola-Bhatti Wildlife Sanctuary, Delhi, India', *Environmental Monitoring and Assessment* 186(9): 5911–20.

Lefebvre, Henri. 1991. *The Production of Space*, Vol. 142. Oxford: Wiley-Blackwell.

Ludden, David. 2011. *The New Cambridge History of India: An Agrarian History of South Asia*, Part 4 Vol. 4. Cambridge: Cambridge University Press.

Lynch, Kevin. 1960. *The Image of the City*, Vol. 11. Cambridge, Massachusetts: MIT Press.

Maheshwari, J. K. 1963. *The Flora of Delhi.* New Delhi: Council of Scientific & Industrial Research, New Delhi.

Mann, Michael, and Samiksha Sehrawat. 2009. 'A City with a View: The Afforestation of the Delhi Ridge, 1883–1913', *Modern Asian Studies* 43 (02): 543–70.

Mishra, A. K., M. P. Sharma, and H. Singh. 2015. 'Floristic Diversity of Delhi, India: A Checklist', *International Journal of Herbal Medicine*, 3(4): 8–18.

Mishra, J. K., R. Aarathi, and M. D. Joshi. 1994. 'Remote Sensing Quantification and Change Detection of Natural Resources over Delhi', *Atmospheric Environment* 28(19): 3131–37.

MoEFCC. 2016. Draft Notification S.O.2114(E), 15 June 2016.

Mohan, Manju, Subhan K. Pathan, Kolli Narendrareddy, et al. 2011. 'Dynamics of Urbanization and its Impact on Land-Use/Land-Cover: A Case Study of Megacity Delhi', *Journal of Environmental Protection* 2(09): 1274.

Montgomery, Mark. 2007. 'United Nations Population Fund: State of World Population 2007: Unleashing the Potential of Urban Growth', *Population and Development Review* 33(3): 639–41.

Naudiyal, Niyati, Joachim Schmerbeck, and Stefanie Gartner. 2017. 'What Influences the Plant Community Composition on Delhi Ridge? The Role Played by *Prosopis juliflora* and Anthropogenic Disturbances', *Tropical Ecology* 58(1): 33–43.

Notification No. F.10(42)-1/PA/DCF/93/2012-17(I).

Order, 12 April 2013.

———. In the matter of land in Gujjar village, Manesar Tehsil.

Page, J. A. 1937. 'A Memoir on Kotla Firoz Shah, Delhi', *Memoirs of the Archaeological Survey of India* 52, Delhi: Manager of Publications.

Parker, R. N. 1920. 'Afforestation of the "Ridge" at Delhi', *Indian Forester* 46(1): 21–28.

Paul, Somajita, and Harini Nagendra. 2015. 'Vegetation Change and Fragmentation in the Mega City of Delhi: Mapping 25 Years of Change', *Applied Geography* 58: 153–66.

Punjab Government. 1884. *Gazetteer of the Delhi District, 1883–84.* Chandigarh: Punjab Government.

Rangarajan, M. 2001. 'The Hunt and the Wilderness in Mughal India'. In M. Rangarajan, *India's Wildlife History: An Introduction*, p. 135. New Delhi: Permanent Black.

Rao, M. S. Rama Mohan, R. N. Adhikari, S. Chittaranjan, et al. 1996. 'Influence of Conservation Measures on Groundwater Regime in a Semiarid Tract of South India', *Agricultural Water Management* 30(3): 301–12.

Robbins, Paul. 1998. 'Paper Forests: Imagining and Deploying Exogenous Ecologies in Arid India', *Geoforum* 29(1): 69–86.

Shahabuddin, Ghazala. 2016. 'Wildlife: Creatures Great and Small'. In *First Garden of the Republic: Nature in the President's Estate*, pp. 126–59. New Delhi: Publications Division, Ministry of Information and Broadcasting, Government of India.

Sharma, Ajay, and R. C. Pal. 2001. 'Afforestation for Reclaiming Degraded Village Common Land: A Case Study', *Biomass and Bioenergy* 20(6): 35–42.

Sharma, Y. D. 1964. 'Delhi and its Neighbourhood'. New Delhi: Archaeological Survey of India.

Sheikh, Shahana, and Ben Mandelkern. 2014. 'The Delhi Development Authority: Accumulation without Development', A Report of the Cities of Delhi Project, New Delhi: Centre for Policy Research.

Sheikh, Shahana, and Subhadra Banda, 2014. 'The Thin Line between Legitimate and Illegal: Regularising Unauthorised Colonies in Delhi', A Report of the Cities of Delhi Project, New Delhi: Centre for Policy Research.

Singh, Upinder. 2006. *Delhi: Ancient History*. New York: Berghahn Books.

Sinha, G. N. 2014. 'An Introduction to the Delhi Ridge'. New Delhi: Department of Forest and Wildlife, Government of National Capital Territory of Delhi.

Trivedi, Mudit. 2009. 'On the Surface Things Appear to Be … (Perspectives on the Archaeology of the Delhi Ridge)'. In Upinder Singh and Nayanjot Lahiri (eds), *Ancient India New Research*, p. 306. New Delhi: Oxford University Press.

Tyagi, Vikrant, Krishan Kumar, and Vinod Kumar Jain. 2006. 'A Study of the Spectral Characteristics of Traffic Noise Attenuation by Vegetation Belts in Delhi', *Applied Acoustics* 67(9): 926–35.

UN. 2016. 'The World's Cities in 2016—Data Booklet', ST/ESA/SER.A/392. Department of Economic and Social Affairs, Population Division.

Varsha. 2014. 'Delhi Ridge'. In G. N. Sinha (ed.), *An Introduction to the Delhi Ridge*. New Delhi: Department of Forest and Wildlife, Government of National Capital Territory of Delhi.

Véron, R. 2006. 'Remaking Urban Environments: The Political Ecology of Air Pollution in Delhi', *Environment and Planning A*, 38(11): 2093–2109.

Vij, Sumit, and Vishal Narain. 2016. 'Land, Water & Power: The Demise of Common Property Resources in Periurban Gurgaon, India', *Land Use Policy* 50: 59–66.

Wells, Andrew. 2007. 'Imperial Hegemony and Colonial Labor', *Rethinking Marxism* 19(2): 180–94.

Whitehead, Judith. 2012. 'John Locke, Accumulation by Dispossession and the Governance of Colonial India', *Journal of Contemporary Asia* 42(1): 1–21.

Zoological Survey of India. 2003. 'Some Selected Fauna of Asola-Bhati Wildlife Sanctuary', *Fauna of Conservation Area No. 16*, p. 70. Kolkata: Director, Zoological Survey of India.

3

Rewilding with the Asiatic Cheetah

Policy and Politics of Wildlife Reintroduction in India*

GHAZALA SHAHABUDDIN

REWILDING AND CARNIVORE REINTRODUCTION

For the last few decades, conservationists have been advocating 'rewilding' of natural habitats, i.e., restoring ecosystems and resurrecting species which may have become extinct in the historic past (see, for example, Soule and Terborgh 1999). Rewilding includes restoration and consolidation of large uninhabited landscapes to a level where these systems become self-sustaining (ibid.). One of the important dimensions of rewilding is the reintroduction of top predators as they are thought to play a crucial role in maintaining ecosystem diversity and function (Terborgh and Estes 2013).[1] Since the early 1900s, predators have been reintroduced to habitats where they have become locally extinct, reached unviable population size or are threatened by low genetic diversity (Seddon et al. 2007; Zimov 2005).

Since 2000, there has been an acceleration in carnivore reintroductions; more so in the developed countries of North America and Europe (Jorgensen 2013; MacDonald 2009). A recent review shows that as many as 42 reintroductions of carnivores

* This chapter was written while I was a Visiting Scholar at the Centre for Advanced Study of India, University of Pennsylvania, USA. I am very grateful for CASI's assistance and financial support. I am also grateful to Rohan D'Souza, K. Sivaramakrishnan and Mahesh Rangarajan for their extensive comments on this chapter.

have been carried out globally since 1900, out of a total of 172 mammalian reintroductions/translocations (Seddon et al. 2005). Relatively better documented reintroductions are those of the grey wolf to Yellowstone National Park, USA in 1995 (Smith and Bangs 2009) and of pumas to Everglades National Park, USA, also in 1995 (Johnson et al. 2010).

Globally, there have been several cases where reintroduced carnivores were established successfully, such as grey wolf to Yellowstone National Park (Smith and Bangs 2009) and African lions within South Africa (Hayward et al. 2007). However, there are a large number of scientific issues that emerge in species reintroduction. For instance, the methods and protocols for species reintroduction are as yet in early, experimental stages so that the rate of success is estimated to be rather low: 30 per cent (MacDonald 2009). In a typical case, there are issues of low genetic diversity and adaptability of the source population, and ecological unsuitability of the target habitat (International Union for Conservation of Nature [IUCN]/SSC 2013). Such concerns become amplified when the reintroduced individuals come from a captive population that are not adapted to fend for themselves and that do not easily form social groups (Jule et al. 2008). Further, the actual costs of reintroduction can be prohibitive due to the need for a high degree of technological inputs and intensive habitat restoration. Continual management is usually required even after a successful population establishment, as long-term responses of carnivores can be unpredictable (Johnson et al. 2010).

Reintroductions of carnivores also come with many socio-economic concerns, which are often overlooked by managers (IUCN/SSC 2013; MacDonald 2009). A particular concern is the effect of carnivores on social welfare and commercial activities such as animal husbandry, forest product collection and agriculture, particularly where people survive on low economic margins. Often there is no collective memory of the species to be reintroduced among the local residents or traditional knowledge to cope with it; this increases the chances of conflicts with people. Such concerns have been raised in developed countries too, such as in debates over reintroduction of even a small-sized predator such as the lynx into the Scottish countryside.[2, 3] Thus, the scientific, social and economic issues in large carnivore reintroductions are much debated (MacDonald 2009).

Rewilding with the Cheetah in India

It is in this context of big cat conservation that I study a recent rewilding proposal in India: the reintroduction of the Asiatic cheetah, a species that went extinct in India in 1967. The cheetah is an iconic large cat species of arid and semi-arid grasslands. It is said to be the swiftest mammal on the planet: a predator highly specialised for taking antelope as prey. The cheetah is seen by ecologists as a useful flagship species for conservation of grasslands, a traditionally neglected ecosystem (Ranjitsinh and Jhala 2010).[4] The cheetah reintroduction plan (hereafter referred to as the cheetah plan) aimed to translocate African cheetahs to Kuno Wildlife Sanctuary in central India.

The cheetah plan sparked a lively and rancorous debate on the *pros* and *cons* of reintroducing a large carnivore that had been missing from the Indian landscape for half-a-century.[5] People feared that it might deepen existing park–people hostilities in India due to the revival of human–wildlife conflicts. Some saw it as merely a pipedream on the part of influential biologists who were inspired by the emerging concept of rewilding in Western countries. Other ecologists felt that the cheetahs would bring much-needed attention to the issue of decline of grasslands and scrub savannahs, ecosystems hitherto largely neglected by both the government and scientists (Ranjitsinh 2017).

This chapter explores the policies and processes surrounding the cheetah reintroduction plan in India, particularly in terms of the relationships among scientists, civil society, local communities and the government. Through the medium of the cheetah, my chapter seeks to answer these questions: what is the role of science and of politics in the rewilding debate in India? How is this manifested in wildlife conservation policy?

This chapter is primarily based on scanning of newspapers, Internet sites, government reports, court judgments and scholarly articles with the aim of tracking the history of, and perspectives on the cheetah reintroduction plan in India. Further, interviews were carried out with government officers and professional biologists who work on carnivore ecology in India.

Carnivore Reintroduction in India

India is one of the first few developing countries where reintroduction of predators was undertaken early on, albeit with mixed results. In 1957, an attempt was made to reintroduce Asiatic lions from Gir Wildlife Sanctuary in Gujarat to Chandraprabha Wildlife Sanctuary in Uttar Pradesh with the translocation of three lions (Divyabhanusinh 2005). Although the number of lions increased to 11 within a span of eight years, the population dwindled after that, and by 1965, none remained. This failure was ascribed to the small area of the sanctuary, and possibly lack of sufficient habitat. Likely, behavioural attributes of lions had not been taken into account while planning the translocation, nor had the high possibility of conflict with the local people been predicted (ibid.).

Much later in the twenty-first century, an opportunity arose for the reintroduction of Bengal tigers. A decision was taken to reintroduce Bengal tigers to Sariska Tiger Reserve, Rajasthan, from where they had disappeared in 2004, a local extinction ascribed mainly to poaching and poor habitat quality (Johnsingh and Madhusudan 2009; Shahabuddin 2010). Thus, from 2008–10, Bengal tigers (two males and three females) were reintroduced to this Reserve with state-of-the-art technical inputs by the Wildlife Institute of India (WII), with mixed results.

Yet another Bengal tiger reintroduction was undertaken soon after, into Panna Tiger Reserve, Madhya Pradesh. Here, too, poachers had taken their toll on wild tiger populations, wiping them out by 2008.[6] Here a total of six tigers were translocated in 2009, which reproduced successfully, according to unofficial reports. Today 40 tigers are reported from Panna Tiger Reserve and are considered to form a viable population.[7]

To better understand the implications of big cat reintroduction in India, it is important to explore India's wildlife policy historically. One of the major pivots of wildlife policy in India since Independence has been establishment of 'inviolate' protected areas (PAs) all over the country in a variety of ecosystems (Greenough 2003). PAs were envisioned as people-free spaces where nature could thrive in the absence of anthropogenic disturbance, enforced by the Wildlife Protection Act, 1972 (Guha 2003). Thus, India's national parks have been marked by the removal of human settlements since pre-Independence period and curbing of local uses such as fuelwood collection, grazing and forest product collection (Rangarajan and

Shahabuddin 2006; Shahabuddin 2010). The restrictions were frequently enforced without the provision of viable livelihood alternatives to the local residents (Shahabuddin 2010). Further, village displacements have tended to be rather inadequate in terms of compensatory payments, financial allocations and support systems (Shahabuddin and Lakshmi 2014). Due to such measures, PAs still experience social hostility and apathy at their peripheries, which is only exacerbated by various types of conflict with wildlife (Guha 2003; Shahabuddin 2010). It is also important to note that despite the large scale of scientific and financial inputs, many PAs have experienced local extinctions and suffered denotification and degradation.

History of the Cheetah in India

To understand the rationale behind cheetah reintroduction, it is necessary to briefly discuss the sequence of events that led to its extirpation from the Indian subcontinent. The cheetah historically occurred throughout the deserts, grasslands and woodland savannahs of Africa, West Asia, Central Asia and India (Eaton 1974).[8] Being specialised for open-chase predation on antelope and deer, cheetahs prey on a diverse range of herbivore species from hares to wildebeest. However, its preferred prey in East Africa consists of medium-sized antelope such as Thomson's gazelle (ibid.: 52). Cheetah individuals are highly dispersed due to their need for following herbivores, and consequently, the home-ranges can be huge, estimated to be 1,651 sq. km in Namibian farmlands (Marker and Mills 2008) and 1,583 sq. km in Algerian Sahara (Belbachir et al. 2015). In the scrub savannah of South Africa, the home-range of a single cheetah group has been estimated to be 126–95 sq. km (Broomhall et al. 2004).

Cheetahs are thought to have originated in Africa from where they naturally dispersed to India just after the Pleistocene (O'Brien 2013).[9] Two races of the cheetah are recognised today—the African (*Acinonyx jubatus jubatus*) and the Asiatic (*Acinonyx jubatus venaticus*); however, genetically cheetahs have been found to be extremely homogeneous (Divyabhanusinh 2006). Today the African subspecies currently occurs in a number of fragmented populations in Africa such as in Namibia, Botswana, Kenya and Tanzania.[10] Iran holds a very small population of the Asiatic subspecies.[11] In India, historical information points to the presence of the cheetah in semi-arid grasslands and scrub savannahs of entire Indian peninsula from the Ganges River southwards to the plateaus of Andhra Pradesh and Karnataka (ibid.: 2).

Globally the cheetah has seen a decline of 90 per cent in the twentieth century, just in Africa (Roff 2012), and today numbers an estimated 10,000 globally.[12] The cheetah was likely to have been naturally rarer than the Bengal tiger or leopard due to its need for a larger home-range, but was not as ecologically versatile as either of them. Also, its unusually high genetic homogeneity as a species manifests in higher infant mortality and reduced fecundity compared to other cat species (Eaton 1974; O'Brien 1994).

As with most extinctions, negative feedbacks from a variety of causes seem to have led to the cheetah's extirpation from India. In medieval times, Mughal emperors hunted antelope with trained cheetah; hundreds were caught from the wild and kept captive (Divyabhanusinh 2006). By the early 1800s, the cheetah was already rare in India due to the considerable rates of capture (Divyabhanusinh 2006; Eaton 1974). During British times, the hunting of cheetah for trophies became prevalent due to the perceived threat to livestock and human settlements (Divyabhanusinh 2006: 88–92). This must have added considerable pressure on the declining Indian population as even small offtakes could cause local extinctions (Rangarajan 1998). The twentieth century saw the accelerated expansion of agriculture and habitation in the plains and plateau regions, leading to decimation of grasslands.[13] Extreme rarity in the early 1900s within India can be surmised by the fact that the African cheetah were imported in small numbers into the country for sport (Craighead and Craighead 1942; Divyabhanusinh 2006: 152); this continued right up to the 1940s, the time of the demise of the princely states.

Reports vary on the date of the last sighting of the cheetah in India. The last record of a cheetah shoot is from 1947 from Korea, Madhya Pradesh. By 1955, the Indian Board for Wildlife accepted the fact of the cheetah's extinction in India.[14] However, a credible report of a live sighting is from the winter of 1967–68 when M. S. Singh Deo sighted a pair between Surguja and Sidhi districts in northern Madhya Pradesh (Divyabhanusinh 2006: 222).

Discussions on the possibility of reintroducing cheetahs to Indian landscapes had started in 1955, when the government recognised that it had disappeared from India. In the same year, the State Wildlife Board, Andhra Pradesh suggested the reintroduction of the cheetah in the state on an experimental basis.[15] The *pros* and *cons* of reintroduction of cheetahs were discussed by the influential naturalist M. Krishnan in a newspaper article in 1965.[16] In 1984, a paper on the status of the cheetah in India for the Ministry of Environment and Forest (MoEF) sparked interest in this issue at the IUCN (Divyabhanusinh 2006). Several other eminent biologists of the time such as Zafar Futehally[17] and M. K. Ranjitsinh, both of whom had a close rapport with the erstwhile Prime Minister Indira Gandhi,[18] recommended this step. The MoEF formally wrote to the Iranian government to request the cheetahs. In addition, there were offers by the Kenyan government to send cheetahs to India.[19] In November 1998, there were reports of a cheetah breeding facility to be set up at Machiya Nature Reserve, Jodhpur[20] and of a proposal by Lalji Singh, director of the Centre for Cellular and Molecular Biology (CCMB), Hyderabad, to attempt cloning of the cheetah.[21] It was clear by now that the government's talks with Iran on this issue had come to nought due to the political disturbance there, and that the cheetahs were likely to be sourced from south-western or eastern Africa.

Several years later, in 2009, there was renewed activity on this front, reportedly triggered by the Wildlife Trust of India (WTI) and its chair, M. K. Ranjitsinh (Ranjitsinh and Divyabhanusinh 2013). Jairam Ramesh, as the Indian Minister of State, MoEF, was sympathetic to this idea. A high-profile meeting was held in Gajner, Rajasthan to further discuss the cheetah plan, involving the WII, WTI, Cheetah Conservation Fund, Namibia, IUCN and several state forest departments. As a result of this meeting, Ramesh approved the recommendation for a detailed survey of seven potential reintroduction sites in four states. The survey was expected to prioritise sites for reintroduction and prepare an action plan.[22] The specifics regarding the source population remained unclear. One year later, in July 2010, the WII and WTI suggested three possible sites for the cheetah reintroduction (Ranjitsinh and Jhala 2010; hereafter referred to as the WII–WTI Report).[23] However, Kuno Wildlife Sanctuary in Madhya Pradesh was finally chosen,[24] given the reportedly large area of habitat available and the expenditure incurred earlier on village displacement from Kuno for the Asiatic Lion (Sharma and Kabra 2007).[25] Once Kuno Wildlife Sanctuary was decided on, the Madhya Pradesh Forest Department prepared an Action Plan for the reintroduction (Madhya Pradesh Forest Department et al. 2011). However, on the exit of Ramesh in 2011, the cheetah plan seems to have got stalled for want of political support (Ranjitsinh 2017).

To complicate matters, a Public Interest Litigation (PIL) was filed by biologists against the cheetah plan in 1995, heard in 2013. The Supreme Court of India stayed the order to reintroduce the cheetah to Kuno Wildlife Sanctuary on the basis that the project was based on inadequate scientific study.[26, 27, 28] On 24 April 2014, the MoEF petitioned the Supreme Court to vacate the stay on cheetah relocation, denying any short-comings.[29] What transpired thereafter is not known, but on 31 January 2015, newspapers reported that the WII had been commissioned to carry out another survey of habitat suitability for the cheetah in Sathyamangalam Forests, Tamil Nadu.[30] Today, the cheetah plan lies dormant within the dusty cabinets of the MoEF, waiting for political interest.

Debates on the Cheetah Plan

The cheetah plan was dogged by controversy and debate from the time the issue became public. While several natural historians and biologists were in favour of the reintroduction, many others advocated care and caution in implementing it. Some others spoke against the idea. State forest departments did not lend their enthusiastic support to the plan, and in most cases, a guarded response was evident.

The first point of controversy was where the cheetahs would be sourced. It stands to reason that the populations that are closest in genetic make-up to the extinct cheetah of India would be used for translocation to maximise chances of adaptation in the translocation site (IUCN/SSC 2013). However, this happened to be the Iranian population at the time (O'Brien 2013), and the Iranian government had turned down the request, given the small number remaining there. The next best source population was identified as the Namibian one, and unofficial talks had started in 2009 when the cheetah plan was revived (Ranjitsinh 2017). Given the overall genetic homogeneity of the species globally, sourcing from East Africa or South West Africa may have been acceptable. But the question remained: would genetic similarity by itself be sufficient for this sensitive species to adapt in another continent? Cheetahs are notorious for being difficult breeders in captivity, and thus, this import could well be unsuccessful. Influential wildlifers such as Valmik Thapar opposed the cheetah plan as they believed that cheetahs did not occur naturally in India; they were 'exotic aliens' (Thapar et al. 2013). In a review of translocations of four species of carnivores within South Africa from 1985–2005, the cheetah populations were least successful (Hayward et al. 2007).

Another important factor to consider is the site for reintroduction, which should ideally be within the historic range of the species. Indigenous range can only be inferred from historical or written records; only where direct evidence is inadequate, suitable habitat with ecological similarities can be used for reintroduction (IUCN/SSC 2013). Given that the scrub savannahs of central India had been an important area for sourcing cheetahs in medieval times (Ranjitsinh 2017), it was likely to serve as a good reintroduction site. Yet ecologists felt that the scrub savannah habitat may have changed considerably since medieval times, a factor that has to be suitably weighed (IUCN/SSC 2013). Grasslands are complex and dynamic ecosystems and the factors that affect and shape plant species and composition are as yet largely unknown in India, as in many parts of the world. While some grasslands are largely fire-maintained, others may be controlled by geological reasons. Tied into this are questions of past anthropogenic influences such as grazing and fire that may have played a role in maintaining plant species composition in the past, and that have now been modified. In addition, there are issues of degradation, soil compaction and invasion by exotics, caused by over-grazing in areas vacated by people.[31] Such factors could lead to unsuitability of the habitat for ungulates and therefore, it is necessary to study them, at the very least.[32] The WII–WTI Report had little discussion of the habitat required, the degree to which it was suitable for cheetahs and the types of habitat management that were necessarily required for improvement. Clearly in-depth studies were first required to initiate any management restoration of the available habitats if the cheetah was to survive here, as suggested by biologist Ravi Chellam:[33]

> We know little about the kinds of grasslands that existed or the mix of prey species that sustained the cheetah. There is

likely to have been considerable change in these systems since 1500s given climate change and extensive changes in land use. Given the lack of such knowledge, how can we satisfactorily restore such habitats, let alone assure the success of the cheetah introduction?

Further, scepticism about the cheetah plan, stemmed from the estimated area needs for a viable population of the species, given the species' high dispersal capabilities and vast home-range in similar habitats. Even early in the debate, naturalist M. Krishnan talked about the difficulties of providing large enough tracts of grassland habitat for the species.[34] The severe shortage of vast unpopulated stretches of scrub savannah or grassland seemed to preclude a comfortable accommodation for the large cat.[35] Several conservationists questioned the cheetah plan for its impracticality, as did Ullas Karanth, eminent biologist with the Wildlife Conservation Society, India:

> To establish a viable population in the wild we would need 10,000 sqkm area essentially free of dogs, goats, sheep etc. We do not have any such area in India. Losses of introduced animals will be high and sustained for a long time … all this is far too expensive and impractical. So first invest and create such a habitat and then think of cheetahs …[36]

Within the government, too, there were many detractors of the plan. Several news agencies reported that state governments appeared reluctant to set aside the amount of land that would be necessary for the cheetah to survive in the wild. For instance, the erstwhile Chief Wildlife Warden, Rajasthan, was even quoted as saying that 'it was not a wise idea.'[37]

It was all to do with the numbers. It was suggested that initially, 27 cheetahs would be moved to the core area of Kuno Wildlife Sanctuary measuring 347 sq. km, from where they were expected to spread out to a larger forested landscape of 3,200 sq. km surrounding (and including) Kuno Wildlife Sanctuary. Eventually the Report envisaged cheetahs populating an area of 6,500 sq. km in the Kuno landscape.

The WII–WTI Report deemed 3,200 sq. km as sufficient for 100 cheetahs, the number to which the initial released population of 27 cheetahs was expected to multiply. However, given research findings in the scrub savannahs of Africa, it was obvious that the Report had under-estimated the area required. Principally this was because quality habitat in Kuno Wildlife Sanctuary was far lower than the South African habitats on which the home-range estimates were based.[38] Further, there was little information presented in the WII–WTI Report to support the presence of sufficient area of unpopulated (but high quality) landscapes around Kuno Wildlife Sanctuary. No reference was made to existing studies of cheetah reintroduction in Africa to understand the behaviour of the cats after reintroduction, and as to how much time would be required for stabilisation of home-ranges.

Another weakness in both the WII–WTI Report and the Action Plan was the lack of a scientific comparison of the source and reintroduction sites in terms of ecological factors. It is a given that reintroductions often take place in sites which are ecologically quite different from the source site in terms of population density, sex ratios, habitat and climatic conditions (Armstrong and Seddon 2007). However, the cheetah plan overlooked this issue and focused only on one factor—prey (herbivore) density. The WII–WTI Report seemed to reduce the complex ecosystem to a simple predator–prey equation.

The approach to the calculation of prey density in Kuno Wildlife Sanctuary, the most crucial determinant of habitat suitability for the cheetah, was also wanting in rigour,[39] Prey density was extrapolated from unpublished data collected five years earlier in 2005 by a PhD student (Banerjee 2005). To this data, an annual growth rate of 5 per cent was applied uniformly to all the existing prey species to come up with numbers for 2010.[40] Interestingly, densities of different prey species were not given separately in the WII–WTI Report. Abundance was reported under four categories: all wild ungulates, lagomorphs (hares), primates (mainly grey langur) and peafowl. An ideal prey assessment should have included, at the minimum, a species-wise breakdown of prey density so that a clear assessment could have been made of habitat suitability for the cheetah, given its preference for a certain size range. There are a few other assessments of prey density that were carried out by WII in 2004–05 and 2006, which are mentioned in the Supreme Court judgment of 2013; but how these affect the suitability of the habitat for the cheetah is not specified in the WII–WTI Report of 2010.[41]

As stated earlier, the cheetah's optimal prey in Kuno would have been the blackbuck and chinkara, based both on their size/ weight as well as their preference of semi-arid grasslands and open scrub savannahs. Of the two best-suited prey species, the chinkara and blackbuck were reported to be present in very small numbers, while the chital, sambar and nilgai were the most abundant (as of 2010–13). The area also harboured roughly 5,000 head of feral cattle, which had been abandoned by the erstwhile inhabitants. It is unclear to what extent the young of larger herbivores such as the chital, sambar, nilgai and feral cattle would be taken by the cheetah: the extent to which non-optimal prey could replace gazelles such as the blackbuck and chinkara.

In carnivore reintroductions, an important risk to consider is that of transmission of parasites and diseases from the source population to the translocation site (Cunningham 1996; IUCN/SSC 2013). Such diseases may cause risk to other animal populations, which may not have earlier exposure to the pathogen. Medical observations and treatment of the cheetah population that would be brought to India was mentioned as a significant step in the Action Plan (Madhya Pradesh Forest Department et al. 2011). Vaccination of all livestock in a buffer area around Kuno was also planned, but there were no estimates of the numbers or the personnel required to carry out this massive task, and how it would be monitored.

Socio-economic Issues in the Cheetah Plan

Apart from the ecological concerns, difficult socio-economic issues surrounded the cheetah plan. One of the reasons Kuno was chosen was that it was populated by 'tribal pastoralists/hunters with low impact', according to the WII–WTI Report. However, Kuno also happened to be the place from where 24 villages (comprising approximately 5,000 people) were displaced and subsequently resettled just outside Kuno Wildlife Sanctuary's boundaries, during 2000–04 (Sharma and Kabra 2007). The displaced people, mainly belonging to the Sahariya tribe, faced problems of livelihood security and impoverishment even several years after displacement (ibid.). Poor quality of allotted land, low financial allocations, poor infrastructure and lack of public consultation contributed to the poor quality of displacement to the extent that poverty, health and employment issues continue to shadow the population for several years afterwards. Such problems have been common in displacement history (Shahabuddin and Lakshmi 2014). Given the bitter experience of displacement, the local people were likely have opposed carnivore reintroduction.[42] This was particularly true as no public outreach programme was carried out during the planning phases of the reintroduction.

Given the critical social issues that could have impacted the success of the cheetah plan, the types of social research reported in the WII–WTI Report were highly inadequate and amateurish. Questions were asked on meat-eating and hunting practices, which were unlikely to reveal real patterns. Interestingly, there were no interviews in the larger target area (of 3,200 sq. km) surrounding Kuno where the cheetahs were expected to disperse after the initial introduction. Field surveyors mainly concentrated on people who lived in villages immediately outside the south-eastern boundary of Kuno.

The proposed introduction of the cheetah would reportedly cause yet another 169 villages to be displaced (from the potential cheetah habitat of 3,200 sq. km area). The people involved were to be compensated with land and other benefits (Ranjitsinh and Jhala 2010). However, it was unclear where such land allotments could be made and whether it would be locally acceptable, given the historical failures of village displacement in the area.

Issues of human–wildlife conflict that could have cropped up between villagers and cheetah in future were also given short shrift in the WII–WTI Report. Some officials admitted that the cheetah plan was rather unrealistic, given the increasing negative interactions between locals and carnivores in different parts of India.[43] Among large cats, cheetahs are considered the least aggressive towards humans; their aggression is limited to situations where their kills are threatened (Eaton 1974: 161). Studies in East Africa show that when there are enough prey, cheetahs do not show interest in domestic livestock such as sheep or goats (ibid.). However, given the patchy densities of prey present in the landscape, biologists A. J. T. Johnsingh[44] and Ullas Karanth contended that cheetahs were likely to feed on small livestock such as sheep, goats, calves and even village dogs.[45]

Given the likelihood of conflicts arising after cheetah reintroduction, a much-needed plan of public outreach and education was not visible in the WII–WTI Report.[46] As Ravi Chellam said:

> We do not have a record of sensibly engaging with local people to prepare them for a novel new carnivore in their midst or to mitigate the adverse impacts on them or involve them in economic activities around the cheetah. Lack of sufficient networking and sensitisation can lead to havoc with the introduction of a big cat.[47]

Instead of engaging with the grievances and possible difficulties of the local people, the planners saw displacement through the offer of compensatory packages as the only possible interface with the affected villages. The scientists appeared to ignore the local political economy, revealed in statements such as: 'The people residing in the forested areas outside Kuno are poor and backward and a good compensation package like the one offered for the core areas of tiger reserves would be irresistible' (Ranjitsinh and Jhala 2010).

The need for social scientists to be a part of the planning process during carnivore reintroduction cannot be over-emphasised (Doubleday 2018; Page et al. 2015). There is a need for systematic threat assessments as well as opportunities in a translocation of this scale (see, for example, Page et al. 2015). In Kuno, the planners seemed to have lost an invaluable opportunity to establish chances for co-existence among the local communities, mainly due to a lack of expertise in understanding human–animal interactions and perceptions. Means of livelihood such as ecotourism and sustainable forest use have been found to be successful in garnering local support for conservation in many parts of the world.[48] However, these issues were hardly broached in Kuno during the planning process. Overall, the depth of public outreach critical for successful reintroduction was given short shrift. In contrast, the reintroduction of the Eurasian beaver in Great Britain took 10 years due to the elaborate public consultations and parliamentary deliberations, in addition to the scientific studies (MacDonald 2009).

Political Imperatives

From the course of events, it is clear that a strong motivation for the cheetah reintroduction plan was a nationalist one; it was clearly seen as a landmark wildlife project for India, which could bring her into a leadership position vis-à-vis wildlife conservation.

The biologists who had personal and professional interests in the grassland ecosystem triggered the plan, which then received an impetus due to the personal interest taken by the erstwhile MoEF. To be seen successfully transporting and reviving a wild population of the 'fastest cat in the world' was a challenge and a matter of patriotic fulfillment. M. K. Ranjitsinh (quoted in Braun 2009), said: 'The return of the cheetah would make India the only country in the world to host six of the world's eight large cats and the only one to have all the large cats of Asia.'[49]

In a similar vein, the MoEF, in its petition to the Supreme Court in 2010, said: 'It (Cheetah) is the only mammalian species to have gone extinct in peninsular India in historical times and bringing it back will have special significance for the national conservation ethic and ethos.'

Yet, in comparison to the Bengal tiger, the cheetah plan had a relatively weaker lobby, politically or otherwise. The misgivings of a number of eminent biologists have been elaborated earlier. Further, the cheetah also had comparatively limited support from the state governments, in comparison to the Union ministry. No carnivore translocation can be carried out without the involvement and approval of the state forest departments.[50] Some senior officers stated that they would find it difficult to allocate additional habitats at the scale required. For instance, the erstwhile Chief Wildlife Warden of the Rajasthan was even quoted as saying that 'it was not a wise idea.'[51] Further, in sites such as Desert National Park, there were competing priorities such as the Great Indian Bustard, which occupies the same landscape.[52, 53] The cheetah was even seen as an 'exotic alien' by key conservationists such as Valmik Thapar, who had earlier been quite instrumental in tiger conservation. While the cheetah was touted as a key flagship species for the grasslands, it was far from clear whether it could be an effective one. Its scrub savannah habitat still remains relatively obscure and unknown to laypersons, seen as being closer to the 'wasteland' or 'degraded forest' in public consciousness.

Seemingly, the reintroduction of the cheetah was being lobbied for by just a handful of biologists who did see the importance of the scrub savannah ecosystem and its marginalisation in the conservation domain. Support from a sympathetic Union Minister pushed the cheetah plan from a dream to an actionable plan. Given that the cheetah plan remains dormant today, it was the set of contrarians who seem to have won, proving that there are many fissures even within the conservationist community.

Conclusion

Big carnivores loom large in public and governmental consciousness, more than any other animal group, given their cultural and economic significance (Quammen 2003). Prominent conservation programmes frequently centre on charismatic carnivores, reminiscent of earlier times in Indian history when such animals were symbols of State power and identity (Divyabhanusinh 2006; Hughes 2013). The cheetah idea captured the imagination of biologists, the government and the wildlife establishment in India, although the cheetah plan did not have unadulterated support among either the public or other interest groups. Yet, political imperatives overshadowed the scientific ones: the government's use of scientific facts and tools for justifying the cheetah reintroduction was selective. Ecological issues of concern were not discussed or solved adequately. Further, the requirements for enabling co-existence with the local populace in future were only superficially explored or even understood. I have shown in this chapter that governments sometimes respond largely to political motivations when they are involved in conservation decision-making; this compromises science as well as true public participation.

Notes

1. See various chapters in Terborgh and Estes, 2013, *Trophic Cascades: Predators, Prey and the Changing Dynamics of Nature* (Washington, D.C., Covelo and London: Island Press).
2. David Miller, 2015, 'Call for Lynx and Wolf Reintroduction', *BBC News*, 15 July 2015, available at http://www.bbc.com/news/uk-scotland-33533035 (accessed on 9 September 2015).
3. Lynx UK Trust, 2015, 'Bringing Back the Lynx to the British Isles', available at http://www.lynxuk.org/ (accessed on 9 September 2015).
4. Jason Burke, 2010, 'India Approves Plan to Reintroduce Cheetah', *The Guardian*, available at https://www.theguardian.com/environment/2010/jul/29/india-cheetah (accessed on 14 September 2015).

5. Milind Pariwakam, 'Why Leave the Cheetah Out?', *Tehelka*, 28 June 2012, available at http://archive.tehelka.com/story_main53.asp?filename=Ws280612Cheetah.asp (accessed on 14 September 2012).
6. Rahul Noronha, 2017, 'Challenges Ahead as Tigers from Madhya Pradesh's Panna National Park Spread Out to Neighbouring Habitats', *Mail Online India*, 1 July 2017, available at http://www.dailymail.co.uk/indiahome/indianews/article-4657780 (accessed on 15 May 2018).
7. Times News Network, 2014, 'Success of Tiger Reintroduction at Panna Tiger Reserve', *The Times of India*, 6 April 2014, available at https://timesofindia.indiatimes.com/city/bhopal/Success-of-tiger-reintroduction-at-Panna-Tiger-Reserve/articleshow/33326120.cms(accessed on 18 May 2018).
8. IUCN Red List of Threatened Species, *Acinonyx jubatus*, available at http://www.iucnredlist.org/details/219/0 (accessed on 23 September 2015).
9. In an influential book, (Valmik Thapar et al. 2013) suggested that cheetahs were exotic to India and had been introduced by anthropogenic means around 300 BC. However, molecular evidence from cheetah remains in Africa and India showed that this contention could not be backed (O'Brien 2013).
10. IUCN Red List of Threatened Species, *Acinonyx jubatus*.
11. Cheetah Conservation Fund, 'About the Cheetah, Race for Survival', available at http://cheetah.org/site/wp-content/uploads/2014/02/Past-and-present-cheetah-range-map.jpg (accessed on 9 October 2015).
12. ibid.
13. M. Krishnan, 1965, 'The Cheetah: Our Wildlife', *The Times of India*, 6 June 1965, p. 8.
14. 'Indian Cheetah, Return to Andhra Jungles Planned', *The Times of India*, 25 November 1955, p. 9.
15. ibid.
16. Krishnan, 'The Cheetah: Our Wildlife'.
17. Tariq Aziz, 1995, 'Debate Rages over Cheetahs' Comeback', *The Times of India*, 1 June 1995, p. 13.
18. ibid.
19. Usha Rai, 1984, 'African Cheetahs for India', *The Times of India*, 27 October 1984, p. 5.
20. *Down to Earth* (Web Edition), 'Return of the Cheetah', 30 November 1998, available at http://www.downtoearth.org.in/news/return-of-the-cheetah-22772 (accessed on 17 September 1998).

21. 'India to Clone Cheetah', *BBC News*, 16 October 2000, available at http://news.bbc.co.uk/2/hi/south_asia/974858.stm (accessed on 17 September 2015).
22. David Braun, 'India Asks for Road Map for Reintroduction of Cheetahs, 13 October 2009, available at http://voices.nationalgeographic.com/2009/10/13/cheetahs_to_return_to_india (accessed on 10 September 2015).
23. M. K. Ranjitsinh and Y. V. Jhala, 2010, 'Assessing the Potential for Reintroducing the Cheetah in India', Dehradun: WII; and NOIDA: WTI.
24. Neha Sinha, 'Plan to Return Cheetah to India Okayed, Three Sites Identified', *The Indian Express*, 29 July 2010, available at http://archive.indianexpress.com/news/plan-to-return-cheetah-to-india-okayed-thre/653280/ (accessed on 17 September 2015).
25. Himanshu Kaushik, 'Gujarat Misses out on Cheetah', *The Times of India*, 29 July 2010, available at http://epaper.timesofindia.com/Default/Scripting/ArticleWin.asp? From=Archive&Source=Page&Skin= TOINEW& BaseHref=TOIA/2010/07/ 29&PageLabel=4& EntityId= Ar00401&ViewMode= HTML&GZ=T (accessed on 17 September 2015).
26. Kumar S. Shrivastava, 'Supreme Court Stalls Centre's Plan to Reintroduce Cheetahs to India', *Down to Earth* (Web Edition), 8 May 2012, available at http://www.downtoearth.org.in/news/supreme-court-stalls-centres-plan-to-reintroduce-cheetahs-in-india-38129 (accessed on 16 September 2015).
27. Priscilla Bajaj, 'Supreme Court Green Light for Shifting of Gir Lions to M.P.', *The Hindu*, 16 April 2013, available at http://www.thehindu.com/todays-paper/supreme-court-green-light-for-shifting-of-gir-lions-to-mp/article4622038.ece (accessed on 10 May 2018).
28. ibid.
29. Utkarsh Anand, 'Govt. Seeks SC Nod for Cheetah Project: Seeks Recall of 2013 Order against its Re-introduction Plan', *The Indian Express*, 23 April 2014, available at http://indianexpress.com/article/india/india-others/govt-seeks-sc-nod-for-cheetah-project/ (accessed on 16 September 2015).
30. Richa Sharma, 'Plan to Let the Cheetah Roam Again in the State', *New Indian Express*, 31 January 2015, available at http://www.newindianexpress.com/nation/Plan-to-Let-the-Cheetah-Roam-Again-in-State/2015/01/31/article2645223.ece 9 (accessed on 17 September 2015).

31. This was visible during my field trip to this site in 2011. Most of the vacated village areas had not recovered to grassland or scrub savannahs due to invasion by agricultural weeds.
32. Based on observations during my field trip to Kuno Wildlife Sanctuary in August 2011.
33. Interview with Ravi Chellam dated 20 August 2015.
34. Krishnan, 'The Cheetah: Our Wildlife'.
35. ibid.
36. Interview with Ullas Karanth dated 13 August 2015.
37. John Platt, 2009, 'Should Cheetahs be Reintroduced in India?', 16 September 2009, available at http://blogs.scientificamerican.com/extinction-countdown/should-cheetahs-be-reintroduced-in-india/ (accessed on 10 September 2017).
38. Observations during my field visit to Kuno Wildlife Sanctuary in August 2011.
39. Supreme Court Judgement 2013, 'Centre for Environmental Law, World Wide Fund-India v. Union of India &Ors., (2013) 8 SCC 234'.
40. This growth rate of herbivore prey is likely to be inaccurate as Kuno habitat was different from that of Gir. Also, different species will exhibit differing growth rates.
41. Supreme Court Judgement 2013.
42. Subsequently, the translocation of lions was repeatedly opposed by the Gujarat government. See Kumar S. Shrivastava, 'Will it be Lions or Cheetahs that will Roam Kuno Wildlife Sanctuary?', *Down to Earth* (Web Edition), 2 April 2012, available at http://www.downtoearth.org.in/news/will-it-be-lions-or-cheetahs-that-will-roam-kunopalpur-sanctuary--37927) (accessed on 16 September 2015).
43. Prachi Bhuchar, 2009, 'Tiger Experts Say No to Cheetahs', *NDTV* (Web Edition), 15 September 2009, available at http://www.ndtv.com/india-news/tiger-experts-say-no-to-cheetahs-401492 (accessed on 15 September 2015).
44. Quoted in M. K. Ranjitsinh, 2010, 'Return of the Cheetah?', *Sanctuary Asia*, October 2010, pp. 56–60.
45. Interview with Ullas Karanth, 8 August 2015.
46. Interview with Ravi Chellam dated 20 August 2015.
47. ibid.
48. See G. Shahabuddin, 2010, *Conservation at the Crossroads: Science, Society and the Future of India's Wildlife* (Ranikhet: Permanent Black) for several examples of successful involvement of local communities in wildlife conservation.

49. Braun, 'India Asks for Road Map for Reintroduction of Cheetahs'.
50. It is interesting to note that both the Sariska and Panna reintroductions involved movement of tigers from other PAs within the respective states. Inter-state translocations seemed well-nigh impossible as was seen in the case of the Gujarat lions.
51. Platt, 'Should Cheetahs be Reintroduced in India?'
52. S. Sebastian, 2009, 'Cheetah to Get New Lease of Life in India, *The Hindu*, 21 September 2009, available at http://www.indiaenvironmentportal.org.in/search/?q=sunny%20sebastian&page=3(accessed on 1 May 2018).
53. Government of Rajasthan, 'Project Great Indian Bustard', available at http://www.rajforest.nic.in/?q=project-great-indian-bustard (accessed on 19 October 2015).

References

Anand, Utkarsh. 'Govt. Seeks SC Nod for Cheetah Project: Seeks Recall of 2013 Order against its Re-introduction Plan', *The Indian Express*, 23 April 2014. Available at http://indianexpress.com/article/india/india-others/govt-seeks-sc-nod-for-cheetah-project/ (accessed 16 September 2015).

Armstrong, D. P., and P. J. Seddon. 2007. 'Directions in Reintroduction Biology', *Trends in Ecology and Evolution* 23(1): 20–25.

Aziz, Tariq. 1995. 'Debate Rages over Cheetahs' Comeback', *The Times of India*, 1 June 1995, p. 13.

Bajaj, Priscilla. 2013. 'Supreme Court Green Light for Shifting of Gir Lions to M.P.', *The Hindu*, 16 April 2013. Available at http://www.thehindu.com/todays-paper/supreme-court-green-light-for-shifting-of-gir-lions-to-mp/article4622038.ece (accessed 10 May 2018).

Banerjee, K. 2005. 'Estimating the Ungulate Abundance and Developing the Habitat-specific Effective Strip Width Models in Kuno Wildlife Sanctuary, Madhya Pradesh', PhD Dissertation, Dehradun: Forest Research Institute, India.

Belbachir, F., N. Pettorelli, T. Wacher, et al. 2015. 'Monitoring Rarity: The Critically Endangered Saharan Cheetah as a Flagship Species for a Threatened Ecosystem', *PLOS ONE* 10(1): e0115136. Available at https://journals.plos.org/plosone/article?id=10.1371/journal.pone.0115136 (accessed 10 May 2018).

Bhuchar, Prachi. 2009. 'Tiger Experts Say No to Cheetahs', *NDTV* (Web Edition), 15 September 2009. Available at http://www.ndtv.com/india-news/tiger-experts-say-no-to-cheetahs-401492 (accessed 15 September 2015).

Broomhall, L. S., M. G. L. Mills, and J. T. du Toit. 2004. 'Home Range and

Habitat Use by Cheetahs (*Acinonyx jubatus*) in the Kruger National Park, *Journal of Zoology* 261(2): 119–28.

Bhuchar, Prachi. 'Tiger Experts Say No to cheetahs', *NDTV* (Web Edition), 15 September 2009. Available at http://www.ndtv.com/india-news/tiger-experts-say-no-to-cheetahs-401492 (accessed 15 September 2015).

Braun, David. 'India Asks for Road Map for Reintroduction of Cheetahs', 13 October 2009. Available at http://voices.nationalgeographic.com/2009/10/13/cheetahs_to_return_to_india (accessed 10 September 2015).

Burke, Jason. 2010. 'India Approves Plan to Reintroduce Cheetah', *The Guardian*. Available at https://www.theguardian.com/environment/2010/jul/29/india-cheetah (accessed 14 September 2015).

Cheetah Conservation Fund. 'About the Cheetah, Race for Survival'. Available at http://cheetah.org/site/wp-content/uploads/2014/02/Past-and-present-cheetah-range-map.jpg (accessed 9 October 2015).

Craighead, J., and F. Craighead, Jr. 1942. 'Life with an Indian Prince: As Guests of the Maharaja's Brother, Two Young American Naturalists Study Age-Old Methods of Hunting with Trained Falcons and Cheetahs and Savor the Pomp of Royal India', *National Geographic* 81(2): 235–72.

Cunningham, A. A. 1996. 'Disease Risks of Wildlife Translocations', *Conservation Biology* 10(2): 349–53.

Divyabhanusinh. 2005. *The Story of Asia's Lions*. Mumbai: Marg Publications.

———. 2006. *The End of a Trail: The Cheetah in India*, Second Edition. Delhi: Oxford University Press.

Doubleday, K. 2018. 'Human–Tiger (Re) Negotiations, A Case Study from Sariska Tiger Reserve, India', *Society and Animals* 26(2): 1–23.

Down to Earth (Web Edition). 'Return of the Cheetah', 30 November 1998. Available at http://www.downtoearth.org.in/news/return-of-the-cheetah-22772 (accessed 17 September 1998).

Eaton, R. L. 1974. *The Cheetah: The Biology, Ecology and Behavior of an Endangered Species*. New York: Van Nostrand Reinhold Company.

Government of Rajasthan. 'Project Great Indian Bustard'. Available at http://www.rajforest.nic.in/?q=project-great-indian-bustard (accessed 19 October 2015).

Greenough, P. 2003. 'Pathogens, Pugmarks and Political "Emergency": The 1970s South Asian Debate on Nature'. In P. Greenough and A. L. Tsing (eds), *Nature in the Global South: Environmental Projects in South and Southeast Asia*, pp. 201–30. Durham and London: Duke University Press.

Guha, R. 2003. 'The Authoritarian Biologist and the Arrogance of Anti-humanism: Wildlife Conservation in the Third World'. In V. K. Saberwal and M. Rangarajan (eds), *Battles over Nature: Science and Politics of Wildlife Conservation*, pp. 139–57. New Delhi: Permanent Black.

Hayward, M. W., J. Adendorff, J. O'Brien, et al. 2007. 'Reintroduction of Large Carnivores to the Eastern Cape, South Africa: An Assessment', *Oryx* 41(2): 205–14.

Hughes, J. E. 2013. *Animal Kingdoms: Hunting, the Environment and Power in the Indian Princely States.* Boston, MA: Harvard University Press.

'Indian Cheetah, Return to Andhra Jungles Planned', *The Times of India,* 25 November 1955, p. 9.

'India to Clone Cheetah', *BBC News,* 16 October 2000. Available at http://news.bbc.co.uk/2/hi/south_asia/974858.stm (accessed 17 September 2015).

IUCN Red List of Threatened Species, *Acinonyx jubatus.* Available at http://www.iucnredlist.org/details/219/0 (accessed 23 September 2015).

IUCN/SSC. 2013. *Guidelines for Reintroductions and Other Conservation Translocations. Version 1.0,* pp. viii, 57. Gland, Switzerland: IUCN Species Survival Commission.

Johnsingh, A. J. T., and M. D. Madhusudan. 2009. 'Tiger Reintroduction in India: Conservation Tool or Costly Dream?' In M. W. Hayward and M. J. Somers (eds), *Reintroduction of Top-Order Predators,* pp. 147–63. West Sussex: Blackwill Publishing.

Johnson, W. E., D. P. Onorato, M. E. Roelke, et al. 2010. 'Genetic Restoration of the Florida Panther', *Science* 329(5999): 1641–45.

Jorgensen, D. 2013. 'Reintroduction and De-Extinction', *Bioscience* 63(9): 719–20.

Jule, K. R., L. A. Leaver, and E. G. Lea. 2008. 'The Effects of Captive Experience on Reintroduction Survival in Carnivores: A Review and Analysis', *Biological Conservation* 141(2): 355–63.

Kaushik, Himanshu. 'Gujarat Misses out on Cheetah', *The Times of India,* 29 July 2010. Available at http://epaper.timesofindia.com/ Default/ Scripting/ArticleWin.asp?From=Archive&Source=Page&Skin= TOINEW&BaseHref=TOIA/2010/07/29&PageLabel= 4&EntityId=Ar00401&ViewMode=HTML&GZ=T (accessed 17 September 2015).

Krishnan, M. 1965. 'The Cheetah: Our Wildlife', *The Times of India,* 6 June 1965, p. 8.

Lynx UK Trust. 2015. 'Bringing Back the Lynx to the British Isles'. Available at http://www.lynxuk.org/ (accessed 9 September 2015).

MacDonald, D. W. 2009. 'Lessons Learnt and Plans Laid: Seven Awkward Questions for the Future of Reintroductions'. In M. W. Hayward and M. J. Somers (eds), *Reintroduction of Top-Order Predators,* pp. 411–48. West Sussex: Blackwell Publishing.

Madhya Pradesh Forest Department, NTCA, WTI and Cheetah Task Force. 2011. *Action Plan for the Reintroduction of the Cheetah (Acinonyx jubatus) in Kuno-Palpur Wildlife Sanctuary.* Madhya Pradesh: MP Forest Department.

Marker, L. L., and M. G. L. Mills. 2008. 'Spatial Ecology of Cheetahs on North-central Namibian Farmlands', *Journal of Zoology* 274(3): 226–38.

Miller, David. 2015. 'Call for Lynx and Wolf Reintroduction', *BBC News,* 15 July 2015. Available at http://www.bbc.com/news/uk-scotland-33533035 (accessed 9 September 2015).

Noronha, Rahul. 2017. 'Challenges Ahead as Tigers from Madhya Pradesh's Panna National Park Spread Out to Neighbouring Habitats', *Mail Online India,* 1 July 2017. Available at http://www.dailymail.co.uk/indiahome/indianews/article-4657780 (accessed 15 May 2018).

O'Brien, S. J. 1994. 'A Role for Molecular Genetics in Biological Conservation', *Proceedings of National Academy of Sciences* 91(13): 5748–55.

———. 2013. 'The Exotic Aliens Controversy: A View from Afar', *Journal of the Bombay Natural History Society* 110(2): 108–13.

Page, S. K., D. M. Parker, D. M. Peinke, et al. 2015. 'Assessing the Potential Threat Landscape of a Proposed Reintroduction Site for Carnivores', *PLOS ONE* 10(3): e0122782.

Pariwakam, Milind. 'Why Leave the Cheetah Out?', *Tehelka,* 28 June 2012. Available at http://archive.tehelka.com/story_main53.asp?filename=Ws280612Cheetah.asp (accessed 14 September 2012).

Platt, John. 2009. 'Should Cheetahs be Reintroduced in India?', 16 September 2009. Available at http://blogs.scientificamerican.com/extinction-countdown/should-cheetahs-be-reintroduced-in-india/ (accessed 10 September 2017).

Quammen, D. 2003. *Monster of God: The Man-Eating Predator in the Jungles of History and the Mind.* New York: W. W. Norton & Company.

Rai, Usha. 1984. 'African Cheetahs for India', *The Times of India,* 27 October 1984, p. 5.

Rangarajan, M. 1998. 'The Role of Administration in Extermination: Fresh Evidence of the Cheetah (*Acinonyx jubatus*) in India', *Journal of the Bombay Natural History Society* 95(2): 328–32.

Rangarajan, M., and G. Shahabuddin. 2006. 'Displacement and Relocation from Protected Areas: Towards a Biological and Historical Synthesis', *Conservation & Society* 4(3): 359–78.

Ranjitsinh, M. K. 2010. 'Return of the Cheetah?', *Sanctuary Asia,* October 2010, pp. 56–60.

———. 2017. *A Life with Wildlife: From Princely India to the Present.* Noida, UP: HarperCollins Publishers.

Ranjitsinh, M. K., and Divyabhanusinh. 2013. 'Lion and Cheetah in India: A Critique', *Journal of the Bombay Natural History Society* 110(1): 3–21.

Ranjitsinh, M. K., and Y. V. Jhala. 2010. 'Assessing the Potential for Reintroducing the Cheetah in India'. Dehradun: WII; and Noida, Uttar Pradesh: WTI.

Reintroduction Biology, *Conservation Biology* 21(2): 303–12.

Roff, S. 2012. 'Cheetahs on the Edge', *National Geographic* 222(5): 110–23.

Sebastian, S. 2009. 'Cheetah to Get New Lease of Life in India', *The Hindu*, 21 September 2009. Available at http://www.indiaenvironmentportal.org.in/search/?q=sunny%20sebastian&page=3 (accessed 1 May 2018).

Seddon, P. J., D. P. Armstrong, and R. F. Maloney. 2007. 'Developing the Science of Reintroduction Biology', *Conservation Biology* 21(2):303-312.

Seddon, P. J., P. S. Soorae, and F. Launay. 2005. 'Taxonomic Bias in Reintroduction Projects', *Animal Conservation* 8(1): 51–58.

Shahabuddin, G. 2010. *Conservation at the Crossroads: Science, Society and the Future of India's Wildlife*. Ranikhet: Permanent Black.

———. 2011. Observations, Field Trip, Kuno Wildlife Sanctuary, 2011.

———. 2015a. 'Written interview with Ullas Karanth', 8 August 2015.

———. 2015b. 'Interview with Ullas Karanth', 13 August 2015.

———. 2015c. 'Interview with Ravi Chellam', 20 August 2015.

Shahabuddin, G., and B. Lakshmi. 2014. 'Conservation-Induced Displacement: Recent Perspectives from India', *Environmental Justice* 7(5): 122–29.

Sharma, A., and A. Kabra. 2007. 'Displacement as a Conservation Tool: Lessons from Kuno Wildlife Sanctuary, Madhya Pradesh'. In G. Shahabuddin and M. Rangarajan (eds), *Making Conservation Work: Towards Innovative Strategies for Securing Biodiversity in this New Century*, pp. 21–47. Delhi: Permanent Black.

Sharma, Richa. 'Plan to Let the Cheetah Roam Again in the State', *New Indian Express*, 31 January 2015. Available at http://www.newindianexpress.com/nation/Plan-to-Let-the-Cheetah-Roam-Again-in-State/2015/01/31/article2645223.ece 9 (accessed 17 September 2015).

Shrivastava, Kumar S. 2012a. 'Will it be Lions or Cheetahs that will Roam Kuno Wildlife Sanctuary?', *Down to Earth* (Web Edition), 2 April 2012. Available at http://www.downtoearth.org.in/news/will-it-be-lions-or-cheetahs-that-will-roam-kunopalpur-sanctuary--37927) (accessed 16 September 2015).

———. 2012b. 'Supreme Court Stalls Centre's Plan to Reintroduce Cheetahs to India', *Down to Earth* (Web Edition), 8 May 2012. Available at http://www.downtoearth.org.in/news/supreme-court-stalls-centres-plan-to-reintroduce-cheetahs-in-india-38129 (accessed 16 September 2015).

Sinha, Neha. 'Plan to Return Cheetah to India Okayed, Three Sites Identified', *The Indian Express*, 29 July 2010. Available at http://archive.indianexpress.com/news/plan-to-return-cheetah-to-india-okayed-thre/653280/ (accessed 17 September 2015).

Smith, D. W., and E. E. Bangs. 2009. 'Reintroduction of Wolves to Yellowstone National Park: History, Values and Ecosystem Restoration'. In M. W.

Hayward and M. J. Somers (eds), *Reintroduction of Top-Order Predators*, pp. 92–125. West Sussex: Blackwell Publishing.

Soule, M. E., and J. Terborgh. 1999. 'The Policy and Science of Regional Conservation'. In M. E. Soule and J. Terborgh (eds), *Continental Conservation: Scientific Foundations of Regional Reserve Networks*, pp. 1–19. Washington, D.C. and Covelo: Island Press.

Supreme Court Judgement. 2013. 'Centre for Environmental Law, World Wide Fund-India v. Union of India &Ors., (2013) 8 SCC 234'.

Terborgh, J., and J. Estes (eds). 2013. *Trophic Cascades: Predators, Prey and the Changing Dynamics of Nature.* Washington, D.C., Covelo and London: Island Press.

Thapar, V., R. Thapar, and Y. Ansari. 2013. *Exotic Aliens: The Lion and the Cheetah in India.* Delhi: Aleph Books.

Times News Network. 2014. 'Success of Tiger Reintroduction at Panna Tiger Reserve', *The Times of India,* 6 April 2014. Available at https://timesofindia.indiatimes.com/city/bhopal/Success-of-tiger-reintroduction-at-Panna-Tiger-Reserve/articleshow/33326120.cms(accessed 18 May 2018).

Zimov, S. A. 2005. 'Pleistocene Park: Return of the Mammoth's Ecosystem', *Science* 308 (5723): 796–98.

4

Threats to Coexistence of Humans and Forests in Central India*

MEGHNA AGARWALA, RUTH DEFRIES, Y. V. JHALA AND Q. QURESHI

INTRODUCTION

The forests of central India provide multiple ecosystem services: they are inhabited by wildlife such as tigers, leopards, the gaur and bears (Sharma et al. 2013; Yumnam et al. 2014); and over 60 per cent of local people are at least partially dependent on these forests for their livelihoods (Saigal 2008). The forests also provide services at regional scales by protecting the headwaters of important rivers, and at the global scale due to carbon sequestration.

It is uncertain whether these forests will continue to provide these services in future, as the forests face multiple threats from mining and infrastructure development (Dutta et al. 2016). The impacts of infrastructure development include increased barriers to wildlife movement that would lead to genetic isolation of wildlife populations and jeopardise the long-term viability of several low-density, wide-ranging wildlife species (ibid.). However, these threats operate at large scales. This chapter focuses on threats to the coexistence of people and forests in central India, which necessarily operate at very local scales. Yet, it is at this local scale

* We would like to thank Ghazala Shahabuddin and K. Sivaramakrishnan for providing us with this opportunity, to present a paper in the workshop on 'Wildlife Conservation in India: From Policy to Practice' and for their comments on an original draft of this chapter. We would like to thank Mahesh Rangarajan and M.D. Madhusudan for their detailed comments. Discussions with Ankila Hiremath, Jayashree Ratnam and M.F. Quamar helped to improve the chapter.

that forests are used and impacted by local inhabitants for their livelihoods. Further, the scale at which we examine the system influences results (Black et al. 2003). Therefore, we conducted a series of studies to examine how people use the forests, what impact this has on the forest, and how this is likely to affect people's livelihoods in future.

Understanding human impacts on forests is complicated by the fact that changes in the forest could also occur due to biophysical variations, as sometimes happens with climate change or stochastic processes. To understand the impact of human action on biophysical systems, it is important to disentangle variations in the system due to natural processes and those due to human action. However, it is often difficult to understand these variations. One major problem in examining a system is that our data may capture only short-term trends (Willis and Birks 2006).

With a long history of human occupation in India, it is often difficult to be certain that 'natural' areas considered to be free of human impacts are indeed truly natural (Heywood and Iriondo 2003). Due to long-term pressure on forests from humans, it is possible that the system has been altered. However, many systems can recover to the original state once the external pressure has been removed. On the other hand, if external pressure (human pressure in this case) is intense, it may have caused the system to cross a tipping point, from which recovery to the original state is not possible (Scheffer et al. 2001). When such shifts across tipping points occur, the capacity of the ecosystem to provide the entire spectrum of ecosystem services can change drastically, and often in a non-linear fashion (Allen et al. 2002). Such systems, perturbed beyond their ecological resilience, may be considered degraded. Resilient and degraded systems may have similar patterns of distribution of species, but the processes underlying the distribution have been altered in the latter. Given India's long history of human–forest interaction, it is difficult to disentangle biophysical impacts from human pressures. To distinguish between these two major drivers of ecosystem change, we conducted a series of studies in the Kanha-Pench landscape in central India to understand how people use the forest, the changes in the forest, and their biophysical drivers, and placed these results within the long-term history of forest change in the region.

Central Indian 'Forests'

In this chapter, we define forests as areas designated as 'forest land' by the forest department, which is not cultivated by local people. We make no judgements about the correct classification of this biome, given recent debates about whether the 'forests' of central India may more accurately be classified as 'savannahs' (Ratnam et al. 2011; 2016). According to this debate, tree–grass coexistence in savannah biomes are maintained due to precipitation regimes (Sankaran et al. 2005), or due to demographic factors such as fire and herbivory in a 'disequilibrium' (Sankaran et al. 2004). In such biomes, savannahs are maintained in arid environments, but the system tends towards woody encroachment in more humid environments unless fire and herbivory are present. Savannahs are now defined as regions with precipitation between 1,000 mm and 2,500 mm where a bimodality of grass cover dominance or tree cover dominance is maintained, depending on the presence of fire or herbivory (Staver et al. 2011), characterised by fire-tolerant and shade-intolerant woody tree species.

Scholars suggest that the source of the fire may be irrelevant in this landscape as anthropogenic fire has long replaced natural fire in almost all ecosystems (Ratnam et al. 2011); and globally, it is almost impossible to uncouple the 'natural' background level fires and their impacts from human-set fires through geological time (Bowman et al. 2011). In tropical savannahs, humans influence seasonality and the spatial pattern of fires, while fires in rainforests are almost always human-set (ibid.). The impact of herbivory is more difficult to ascertain, but overgrazing reduced fire historically (1870 CE) (Bowman et al. 2009), reduces fuel-load for light burning (Pyne 2001), and maintains the savannah system in the absence of fire (Sankaran et al. 2013).

Pollen studies show the changing forest composition in central India from 12,700 years ago to present (Chauhan and Quamar 2010; 2012a; 2012b; Chauhan et al. 2013; Quamar and Chauhan 2012; 2014; 2015), when forest composition transitioned between tree savannahs, open deciduous forests and mixed deciduous forests, depending upon climatic conditions and the southwest monsoon. Because other studies have attributed the presence of open forests in South Asia to fires based on charcoal presence in the sediment layer (Morrison 1994), it is possible that fires were also responsible for open forests in central India.

At present, the 'forests' of Central India are tropical deciduous, a classification that is characterised by leaf-fall in the summer months. These forests can be sal-dominated (*Shorea robusta*), teak-dominated (*Tectona grandis*) or mixed forests dominated by *Terminalia*, *Anogeissus* and *Lagerstroemia* species. The forests also contain other species and, are highly heterogeneous in terms of flora and forest structure, and have a relatively open canopy. Species composition often changes every fifty metres.

Historically, the distribution of the forests was not the same as it is now, or even when the British colonised India. Historians have documented a shifting forest frontier, in which forests expanded when conditions were favourable (agrarian taxes were high) and retreated when conditions were unfavourable (famines, wars and low taxes) (Guha 1999: 37, 50, 127; Rangarajan 1996: 36–42; Sivaramakrishnan 1999; Skaria 1998). The forest that regenerated was also very different from the previous forests: historians record that the regenerating forests were a mass of bushes (Guha 1999: 38), had high understory and less tree growth (Prasad 2003: 56–57), or consisted of *Mimosa* and *Acacia* genera (Guha 1999: 38). They report that these kinds of bushes and thorn forests had to be removed, or else the forest would remain in this state (ibid.: 39). This suggests that the forest formed due to human use at different times was different from a hypothetical 'natural' forest.

When the British colonised India, they often notified non-cultivated areas as 'forest'. They often managed these forests as plantations, where they preferentially allowed the growth of certain species and individuals that were tall, had a straight trunk and no branching. Sal trees that grew with saleh (*Boswellia thuriferra*), saj (*Terminalia alata*) and awla (*Phyllanthus emblica*) were eased out in preference for strip planting of what was essentially monoculture sal plantations due to a higher regeneration of sal in the monoculture. This made the forests less able to resist swidden agriculture due to lower soil retention (Prasad 2003: 54–56). The distribution of teak forests in the area may also be a result of teak planting. Specifically, teak timber was planted in the Nagpur region in 1813 (Guha 1999: 125). Therefore, present-day distribution of sal and teak stands may be largely natural, but they may partly be the product of human intervention in the past.

People and Their Forest Use

The Kanha-Pench landscape straddles the districts of Mandla, Seoni and Balaghat in Madhya Pradesh, and includes four forest divisions (East and West Mandla Divisions, North Balaghat Division, South Seoni Division) (Figure 4.1). The area has a high proportion of Scheduled Tribes (STs), with this proportion increasing from Balaghat (21.80 per cent) to Seoni (36.78 per cent) to Mandla (57.23 per cent), although rural percentages are higher: Balaghat (22.85 per cent), Mandla (62.48 per cent) and Seoni (40.14 per cent) (Census of India 2011b). Population densities in the area are lower than in the rest of India (133 people/sq. km in Seoni, 154 people/sq. km in Mandla and 162 people/sq. km in Balaghat; 384 people/sq. km average for India), but have increased in the last 10 years (157 people/sq. km in Seoni, 182 people/sq. km in Mandla and 184 people/sq. km in Balaghat) (Census of India 2011a). The people of the area are mostly small-holders (in 2011, 43 per cent of agricultural land-holdings were less than 1 hectare, and 72 per cent were 1–2 hectares), and they are dependent on surface canals and shallow tanks for irrigation, although access to groundwater is increasing, particularly in Seoni (Mondal et al. 2014). Crops

Figure 4.1: Study Area in Kanha-Pench Landscape in Central India

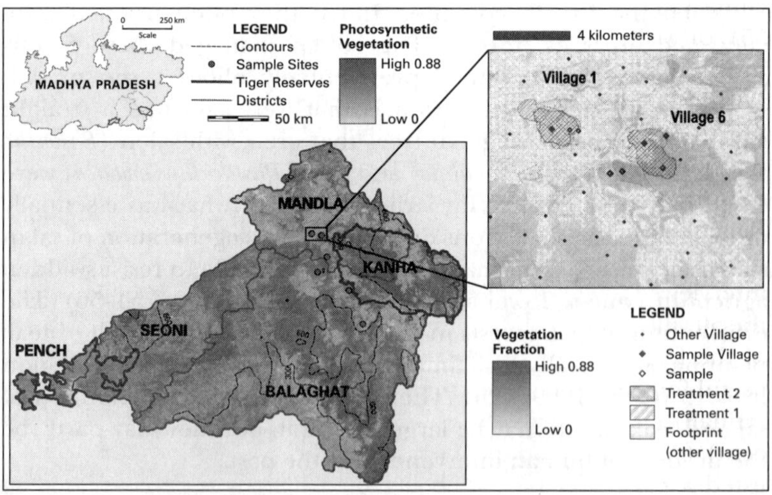

Note: Inset shows forest use surrounding two sample villages.

Source: Agarwala et al. 2016a.

include rice during the monsoon season, and lentils such as *masoor* and *moong* during winter (ibid.). Farmers sell surplus production in the open market or through the government's public distribution system (PDS) (DeFries et al. 2016).

The census also lists a large proportion of people in these districts as non-working (64 per cent in Mandla, 67 per cent in Seoni and 67 per cent in Balaghat), and these 'non-working' people are more likely to be dependent on forests for their livelihoods and needs. The annual value of the top ten commercial non-timber forest products (NTFPs) in the state is estimated to be Rs 3.13 billion, and a large population is dependent on NTFPs for their livelihoods: six million people are involved in *tendu* leaf collection in the state and NTFP collection supports 70 per cent of the household income in forest-fringe villages (Saigal 2008).

The people in the Kanha-Pench landscape also use the forest for fuelwood, house construction and production of local implements. They use the forest to graze their cattle, and less frequently, to graze goats and sheep. People burn the forest to augment both production of NTFPs and fodder for cattle, while forest departments in India invest significant resources in managing fires (Madhya Pradesh Forest Department 2015). This policy has led to conflicts with forest-using communities who either use fires traditionally to manage forests (Gadgil and Guha 1993: 150–58; Hiremath and Sundaram 2013; Prasad 2003: 55), or wish to do so now as they claim it reduces invasive species in the forest (Thekaekara 2014). Local communities across the world have historically used fires to manage the landscape they live in (Preece 2013), and there are reports from central India that people historically used fires to prevent the encroachment of forests, and during shifting cultivation (Prasad 2003: 55; Rangarajan 1996).

To further understand how people use the forest, we interviewed 250 people across the landscape (5 people in each of the 50 villages) about what they considered the most important use of the forest. Equal proportions responded that the forests were useful for their services to the environment (such as clean air and rainfall) (13 per cent) and that the forest had no use at all (13 per cent) (Agarwala et al. 2016b). A significant proportion, however, listed a particular utilitarian use of the forest or named species from the forest (Figure 4.2[a]; Figure 4.2[b]).

Figure 4.2: (a) People's Responses to Questions on the Uses of the Forest; (b) Percentage of Sites where a Species was Browsed

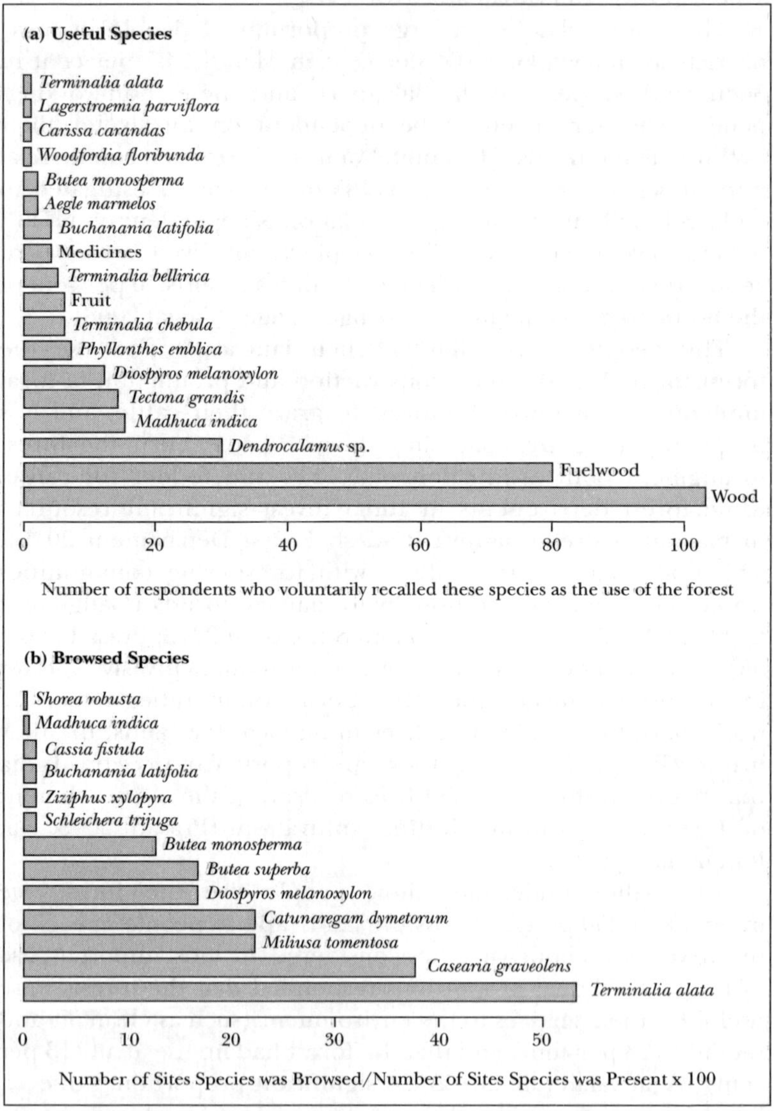

Source: Agarwala et al. 2016b.

Most people thought the forest was important for its *satkatha* or mixed trees that were used for local construction. The next most common use of the forest was *jalau* or its use as fuelwood. In recalling this, many people used the word *nistaar,* a term used in the area for forests that are meant to be used by people to meet their daily needs. Some species named by people included: bamboo (*Dendrocalamus* sp.) that is useful for the construction of homes, particularly for roofing; *mahua* (*Madhuca indica*) that is used for alcohol, oil, and is also sold as an NTFP in the market; teak (*Tectona grandis*) that is used for timber but is not legally allowed to be harvested by local people; *tendu* (*Diospyros melanoxylon*) that is sold as an NTFP in the market; *awla* (*Phyllanthus emblica*), used for food and medicinal purposes and sold in the market (one component of *triphala choorn*); *harra* (*Terminalia chebula*), sold in the market (second component of *triphala choorn*); and *behera* (*Terminalia bellerica*), sold in the market (last component of *triphala choorn*) (for a complete list of species and their uses, see Table 4.1).

Table 4.1: Uses of Tree Species

Species	*Use: Fruit and Flower*	*Use: Leaf and Wood*	*Use: Bark and Sap*
Anogeissus latifolia	None	Leaf: tanning Wood: charcoal, agricultural implements	None
Buchanania latifolia	Fruit: eaten, trade	Leaf: plates Wood: light construction	Bark: tanning Sap: pellucid gum
Butea monosperma	Fruit: Seed purgative, medicine Flower: Yellow dye, medicine	Leaf: lac, plates, fodder for buffaloes Wood: well kerbs	Bark: None Sap: astringent gum
Casearia graveolens	Fruit: poison	Leaf: browse	None

Table 4.1 Contd

Table 4.1 Contd

Species	***Use: Fruit and Flower***	***Use: Leaf and Wood***	***Use: Bark and Sap***
Cassia fistula	Fruit: purgative, medicine	Leaf: cattle fodder (in UP) Wood: agricultural implements	Bark: tanning, dyeing Sap: gum
Dendrocalamus species	None	Leaf: None Wood: roofing, baskets, market	None
Diospyros melanoxylon	Fruit: eaten Flower: None	Leaf: commerce	None
Garuga pinnata	Fruit: eaten, pickled	Leaf: fodder Wood: fuel, indoor work	Bark: tanning Sap: used
Lagerstroemia parviflora	None	Leaf: tanning Wood: agricultural implements, construction	Bark: tanning Sap: sweet gum eaten
Miliusa tomentosa	None	Leaf: cattle fodder Wood: sheds, lopping	None
Madhuca indica	Fruit: oil, poison, insecticide, emetic Flower: alcohol, food	Leaf: None Wood: protected	Bark: None Sap: gum
Phyllanthus emblica	Fruit: food, dyeing, tanning, medicine Flower: None	Leaf: None Wood: water purification, well-work	Bark: tanning Sap: None

Table 4.1 Contd

Table 4.1 Contd

Species	***Use: Fruit and Flower***	***Use: Leaf and Wood***	***Use: Bark and Sap***
Schleichera trijuga	Fruit: oil Flower: None	Leaf: lac, cattle-fodder Wood: crushers, ploughs, etc. (hard wood)	None
Shorea robusta	Fruit: Seed sold as NTFP Flower: None	Leaf: None Wood: Timber	Bark: None Sap: Resin
Terminalia alata	None	Leaf: silkworm, lac pollarded Wood: fuelwood, potash, heavy construction	Bark: tanning, chewed with betel leaf Sap: None
Terminalia chebula	Fruit: commerce, tanning, dyeing, medicine Flower: None	Leaf: galls-ink, dye Wood: furniture, agricultural implements, house building	Bark: tanning and dyeing Sap: None
Ziziphus xylopyrus	Fruit: dye for trade Flower: None	Leaf: fodder Wood: torches, cart-building	Bark: tanning Sap: None

Source: Brandis 2007.

We also gave GPS trackers to people in six villages to document their movement, and the movement of their cattle in the forests in 2012. This required the development of trust, and we had to use the social and kinship networks of our local colleagues to ensure participation (we will describe village selection and sampling in detail later in this chapter).

We found that people used the forest differently across seasons and for different purposes. People minimised effort when collecting fuelwood as it was only legal to carry fuelwood as a head-load. At such times, they travelled an average distance of 1.06 kilometres into the forest. People expended much more effort when there were high value non-timber forest products (NTFPs) in the forest in April–May. During these times, they travelled as much as 3.85 kilometres into the forest to collect these products (Agarwala et al. 2016b). They also had to travel further if there were more people in the village. They also travelled further if their village was located at greater distances from the town (the model that best explained distance travelled by villagers included human population: beta-coefficient .21; distance to town: beta-coefficient 0.18). This could be a function of available livelihood options, as the proportion of 'non-employed' people also increased at greater distances to town, and people in more remote villages had fewer livelihood options and were more dependent on the forest.

Cattle used the forest differently; when there were crops in the fields, they were herded by shepherds to prevent them from destroying agricultural crops. In other seasons, cattle roamed freely in the forest up to a distance of 2.8 kilometres. Cattle moved in different directions in different seasons, and moved further into the forest where there was more forest area around the village (the model that best explained distance travelled by cattle included forest area: beta-coefficient 0.1) (Agarwala et al. 2016b).

They also did not graze indiscriminately, but showed preferences for certain species (Figure 4.2). Cattle most frequently browsed *Terminalia alata, Casearia graveolens, Miliusa tomentosa* and *Catunaregam dumetorum* (Figure 4.2). Of these, *Terminalia alata* was the most abundant species in the forest (greater than 25 individuals in 65 per cent plots); and other species were also common (*Casearia graveolens* and *Diospyros melanoxylon*: greater than 25 individuals in 50 per cent plots), but *M. tomentosa* (greater than 25 individuals in one plot) and *C. dumetorum* were less common (no plots had greater than 25 individual plants) (Agarwala et al. 2016a).

Overall, we find that forest use patterns vary seasonally, and depend on resource availability: people may travel further when commercially viable NTFPs are in season and cattle may travel further where more forest is available. People and cattle are also specific about the species they use.

Studying the Impact of Human Use

A treatment is a site or situation where the factor whose impact we are testing is measured along with its impact, while that factor is absent in the control site. In understanding the impact of human activities on forests in central India, it is difficult to find forests where, historically, there have been no human activities to serve as our control forests. This situation makes it difficult to establish baselines, rates of change, and assess when a forest may be considered 'degraded'. This leaves researchers with only the option of using forests that are used less frequently or those that have no recorded incidence of use, as controls; and such control plots need to be situated close to treatment plots so that there are fewer differences in species pool, biophysical factors such as climate and topography, and management factors such as silviculture. Ideally, only the frequency of use should be different in control plots and treatment plots.

Possible study villages were characterised in terms of population density, livestock density, distance to major towns, and available forest in a 2-kilometre radius, a process that gave three clusters of villages that were similar with respect to these variables. We then selected two villages from each of the three largest clusters for intensive study, i.e., a total of six villages. Table 4.2 shows the values for these parameters for each of the six villages.

Table 4.2: Details of Villages Selected

Village Cluster	*Village ID*	*Total Population*	*Cattle +Buffalo*	*Dist. to Town (km)*	*Forest Cover (km^2 within 2 km radius)*
DH	Village 1	1,007	500	7	1.75
DH	Village 6	846	519	12	4.53
HA	Village 5	881	373	15	2.40
HA	Village 3	781	588	14	4.42
DU	Village 2	403	737	21	7.94
DU	Village 4	755	670	41	8.41

Source: Agarwala et al. 2016b. Based on data collected by Wildlife Institute of India (Total Population, Distance to Town), Department of Animal Husbandry (Cattle and Buffalo populations) and GIS.

Because we tracked forest use around each of these six villages, we could designate areas with high frequency of use (used by both cattle and people in both seasons), intermediate frequency of use (used by either cattle or people in one season), and low frequency of use (no documented use of forest) (for example, see inset in Figure 4.1). We could then compare forests at different frequencies of use in the three different clusters. To avoid the influence of forest department management activities, we selected villages with minimal silvicultural activities in their forest compartments, and selected plots that had no silvicultural activity in their compartment history (Agarwala et al. 2016b).

In the end, we had 20 sampling plots in six villages, wherein each cluster had plots in high, medium and low frequency (of use) areas, where we measured many ecological parameters, including size-class distribution of species, soil compaction, pH and nutrients, canopy cover, understory biomass and signs of human use. We used multiple methods to analyse the data that are detailed in Agarwala et al. 2016a; and 2016b).

Forest Change and Their Drivers

We examined whether site characteristics (for example, population density, livestock density and distance to market), types of human use (for example, species part used, whether species is used consumptively, and whether a species is used for subsistence or for commercial purposes), and plant traits (for example, shade tolerance and fire resistance) were associated with the transition of species from a lower size class to a higher size class. Here, the transition of species from lower size class was quantified as the proportion of individuals in a higher size class relative to the number of individuals in a lower size class in a plot—this indicates the ability of species in a plot to transition to a higher size class if they are present in the lower size class. The rationale behind this approach was that significantly different transition values at sites used at a higher frequency than the controls would suggest that human use is altering relative abundances of the species in the higher size class, although causality can only be inferred (Agarwala et al. 2016a).

Our study found that species that were fire-resistant were more likely to grow from sapling stage to small trees (less than 4 centimetres DBH), but the transition for these species was less likely at higher livestock densities. On the other hand, species with short trunks, which makes them less suitable for local construction, were more likely to grow into large trees (greater than 10 centimetres

DBH); and species that were used for local construction were less likely to grow from medium to large trees (see Table 4.3). Of the drivers identified, those that could impact transition from saplings to small trees are most important, as they impact the reproductive potential of the species by altering the likelihood of individuals reaching adult reproductive stage. These could have a disproportionate impact on future forest composition.

Table 4.3: β-Coefficient Estimates for Factors Associated with Change in Size Class Transition

	Factor	*Estimate*	*Upper Confidence Limit*	*Lower Confidence Limit*
From saplings to small trees (<4 cm DBH)	Fire Resistance	0.14	0.09	0.19
	Livestock per forest area	–0.72	–0.82	–0.62
	Elevation	–0.61	–0.76	–0.46
From medium-sized trees (4-10 cm DBH) to large trees (>10 cm DBH)	Form (tall tree, straight, short trunk)	0.32	0.21	0.43
	Use (local construction)	–0.08	–0.1	–0.12
	Use (sap)	–0.12	–0.13	–0.14
	Population per forest area	0.04	0.036	0.044

Source: Agarwala et al. 2016a.

We also quantified whether relative abundance of different species was changing at higher frequencies of use and at higher population pressure. Because population density was the best predictor for decrease in biomass, understory, tree density and canopy cover (Agarwala et al. 2016b), we expected it to have a strong effect on species distribution as well.

Four species were documented to be increasing at higher frequencies of use (Table 4.4). One species to show an increase, *mahua* (*Madhuca indica*), is the second most recalled species by the people. It is culturally an important species in the landscape and has multiple uses: its flowers are used to make alcohol, and used for sale as an NTFP, and its seeds are used to make oil. Increasing market value for the dried flower means that culturally important products such as *mahua* butter and *mahua* cream have reduced, in preference to its sale as an NTFP. Due to its importance, local people report *de facto* privatisation of *mahua* trees, where individual trees are divided among families, who are then responsible for maintaining it and harvesting their products. Management options include setting small fires below the tree to clear the understory and ripen the flowers. This improves harvest and access as ripe flowers fall to the ground and are easily collected.

Another species to show an increase, *Diospyros melanoxylon* or *tendu*, is also an important NTFP species, and forms an important source of income for the local people where it is sold to the *beedi* (country cigarettes) industry. Six million people are involved in *tendu* leaf collection in Madhya Pradesh, and the total turnover of the *tendu* trade in the state is estimated to be approximately Rs 1.845 billion (Saigal 2008). People also manage forests for *tendu* production, such as by burning the understory to augment leaf production.

Of the two other species that show an increase in relative abundance, *Lagerstroemia parviflora* is resistant to both cattle and fire, while *Butea monosperma* is resistant to cattle. All four species that increased were considered important by people, even if three of them (*Madhuca indica, Butea monosperma* and *Diospyros melanoxylon*) were also browsed by cattle. Such changes have been observed in other forests where intense management by humans leads to retention of only a few species in each forest type, either through selective removal or active management for useful species (Crook and Clapp 1998).

But human action does not only have intentional impacts. Almost all site conditions that were included in the final models (see Table 4.4: livestock and human populations, frequency of use) are anthropogenic, which suggests that alteration of site conditions due to human management was also responsible for changes in species composition (Agarwala et al. 2016a). Such changes may cause unintentional changes in the forest as human activities can create disturbances and species that are resistant to human use may

Table 4.4: Species Traits and Size Class Transitions

Species	*Usefulness Ranking (From Figure 4.2)*	*Traits*	*Statistically Significant Difference in Relative Abundance as Saplings at Higher Frequencies of Use**	*Statistically Significant Difference in Relative Abundance as Small Trees at Higher Frequencies of Use**	*Statistically Significant Difference in Relative Abundance as Medium-sized Trees at Higher Frequencies of Use**
Anogeissus latifolia		Fire-resistant Trampling resistant			
Buchanania latifolia	8	Trampling resistant	–		
Butea monosperma	10	Trampling resistant	+	+	
Casearia graveolens			Complicated		
Cassia fistula		Trampling resistant		–	
Dendrocalamus species	1		–		
Diospyros melanoxylon	4	Trampling resistant	+	+	
Garuga pinnata		Fire-resistant	–	–	
Lagerstroemia parviflora	13	Fire-resistant Trampling resistant	+	+	+

Table 4.4 Contd

Table 4.4 Contd

Species	*Usefulness Ranking (From Figure 4.2)*	*Traits*	*Statistically Significant Difference in Relative Abundance as Saplings at Higher Frequencies of Use**	*Statistically Significant Difference in Relative Abundance as Small Trees at Higher Frequencies of Use**	*Statistically Significant Difference in Relative Abundance as Medium-sized Trees at Higher Frequencies of Use**
Miliusa tomentosa			–	–	
Madhuca indica	2	Not trampling resistant	+		
Phyllanthus emblica	5	Fire-resistant	Complicated		
Schleichera trijuga		Trampling resistant			
Shorea robusta		Fire-resistant	–		
Terminalia alata	14	Somewhat fire-resistant Not trampling resistant at high intensity			
Terminalia chebula	6	Fire-resistant	Complicated		
Ziziphus xylopyrus		Fire-resistant Not trampling resistant			

Note: *(+) Relative abundance was higher at higher frequencies of use; (–) relative abundance was lower at higher frequencies of use.

Source: Compiled from Agarwala et al., 2016b; Brandis 2007; Troup 1983.

increase their proportion in the forest. For example, three of the five species that showed lower relative abundance were grazed by cattle (*Miliusa tomentosa, Shorea robusta* and *Cassia fistula*). Over the long term, other species may further reduce in frequency, become extinct locally and lead to shortages for local people. For example, people reported a shortage of bamboo despite it being the most frequently recalled species, but this reported shortage may also be due to the 12-year flowering cycle of bamboo which leads to the simultaneous death of bamboo over large areas.

Unintentional changes could also reduce species and ecosystem diversity (Crook and Clapp 1998). Several species that were not considered important by local people decreased (*Garuga pinnata, Miliusa tomentosa, Shorea robusta* and *Cassia fistula*) or had other complicated changes (*Bombax ceiba, Bridelia retusa, Schleichera trijuga, Anogeissus latifolia* and *Ziziphus xylopyra*). Further, of those that are frequently present in the forest (Troup 1983), at least two forest species (*Bombax ceiba* and *Boswellia thuriferra*) were absent as small trees, suggesting that these may not be regenerating (although this could be a function of sampling effort) (Agarwala et al. 2016b).

Unintentional Drivers of Change

While active management by people can lead to increase in desirable species (as described in the previous section), in this section, we examine the potential impact of other factors on future forest composition.

Fires

Our studies found that fires led to changes in forest composition due to increased transition to higher size class of fire-resistant species in comparison with other species (Agarwala et al. 2016a). This could potentially lead to an increase in fire-resistant species (see Table 4.4 for this list). This is desirable as three of these are considered important by people (*Phyllanthus emblica, Terminalia chebula* and *Lagerstroemia parviflora*). However, it could also lead to a decrease in species not resistant to fire (*Tectona grandis, Pterocarpus marsupium, Chloroxylon sweitenia* and *Bombax ceiba*), or only somewhat resistant to fire (*Terminalia alata*).

Yet, fires are not necessarily harmful for an ecosystem and the system may be resilient to fires. Globally, palaeoenvironmental studies have established the importance of fire in the biophysical functioning of some forest ecosystems (Burrows 2008; Keane et al. 2008) or how the ecosystem had been formed by a very long history of human-induced fires (Bowman et al. 2011). Such studies have led to changes in the fire management strategy in countries which previously had very strong fire prevention policies (for example, America and Australia), where the countries either abandoned this policy, or now prevent fires less frequently (Allen et al. 2002; Keeley 2006). In central India, too, fire-resistant species are abundant across geological time, although there are some variations when the climatic conditions are drier and there is a weakening monsoon ; but charcoal remains that represent fire in the sediment layer are also present (Chauhan and Quamar 2010; Chauhan et al. 2013). This suggests that fires may play an important role in ecosystem functioning in these forests.

On the other hand, while fire intervals of 5 years increase tree height, basal area and stocking density, annual fires cannot sustain tree biomass (Furley et al. 2008; Ryan and Williams 2011). The current fire regime of annual low intensity fires allow very few species to regenerate and could lead to a large decline in species diversity (Saha and Howe 2003), and may potentially reduce the relative abundance of species such as *Tectona grandis, Terminalia alata, Pterocarpus marsupium, Chloroxylon sweitenia* and *Bombax ceiba,* of which two are considered important by local people and the others may perform important ecosystem services. Studies to establish optimal fire regimes that would best maintain these ecosystems are required (Ratnam et al. 2016).

Human Pressure

Our studies found that human pressure was also significantly associated with forest change. Population density was associated with changing forest structure, frequency of use was associated with changing relative abundance of species, and pressure from grazing influenced the transition of species to higher size classes; such effects could potentially change future species composition of the forest. If we focus on species that are likely to be impacted by livestock (*Madhuca indica, Pterocarpus marsupium, Haldinia cordifolia,*

Ziziphus xylopyrus and *Bombax ceiba* at any grazing intensity, and *Terminalia alata* at intermediate grazing intensity), the list includes species that are considered important by local people (*Madhuca indica* and *Terminalia alata*), and those that are important food sources for animals (*Ziziphus xylopyrus* and *Terminalia alata*). Species that are not impacted include species that are considered important (*Lagerstroemia parviflora, Butea monosperma, Buchanania latifolia* and *Diospyros melanoxylon*), those that are eaten by wildlife (*Schleichera trijuga, Buchanania latifolia, Cassia fistula* and *Diospyros melanoxylon*), invasive species such as *Lantana camara* and other species (*Anogeissus latifolia, Bridelia retusa* and *Chloroxylon sweitenia*).

The history of forest use in the landscape suggests that there have been long-term interaction of humans in this environment, and the system should be adapted to or constructed by human action. While the palynological record provides some evidence for the historicity of fire, there is no concomitant abundance of cattle-resistant species. Many scholars have laid the blame for contemporary forest change on market pressure (Houghton 2012; Olander et al. 2008; Rudel et al. 2009). Yet, the species most valued for their economic value in the market by the local people are not doing badly: six appear to be increasing (*Madhuca indica, Diospyros melanoxylon, Butea monosperma, Lagerstroemia parviflora*) or showing patterns similar to an increase (*Phyllanthus emblica* and *Terminalia chebula*); while only one is decreasing (*Dendrocalamus* spp.).

In fact, our study found that other than bamboo, species most likely to decrease in future were not explicitly valued by local people (*Garuga pinnata, Miliusa tomentosa, Shorea robusta* and *Cassia fistula*). Of these, it is possible that sal is not regenerating because it is in forests that are predominantly sal monoculture, while it requires other mixed species (*Boswellia thuriferra, Phyllanthus emblica* and *Terminalia alata*) to retain soil and regenerate. While sal is not used by local people, the other species may be casualties of being undifferentiated forest species or *satkatha*, where they can be used for multiple purposes, but are replaceable (unlike the species that are recalled, which have specific, unique uses). *Garuga pinnata* is a tall tree with a tall straight trunk, which makes it suitable for local construction. It is also used as fodder for goats, bark, sap, fuelwood, and wood. *Miliusa tomentosa* is used as fodder for cows, and for wood for local construction and furniture. *Cassia fistula* is used as fodder for both goats and cows, and for bark, sap, wood and fuelwood.

All three species are edible, particularly *Miliusa tomentosa*, which is disproportionately popular with animals (Figure 4.2[b]). The loss of these species may have unfortunate consequences for wild animals. Although they can withstand grazing pressure (Agarwala et al. 2016a), it is livestock density that influences transition here—and higher cattle densities may be responsible for reduction in these species that could usually withstand grazing pressure.

Overall, local human pressure was significantly associated with forest change, with greater impacts at greater population densities (human and cattle) and frequencies of use. Such a result may be possible because historic population densities were much lower than they are at present. Historically, forest use was also sometimes restricted: people were taxed by the government for wood and grazing cattle in the forest (Rangarajan 1996: 39–40; 51). The lack of cattle-resistant species in the palynological record also suggests that grazing pressure may not have been high enough to impact abundance. The extent of human pressure in treatment sites today may thus have exceeded historic pressures on the forest, resulting in significant changes in the forest ecosystem. Therefore, even if the forest ecosystem of central India was adapted to or constructed by historic human pressure, high population densities at present may still be unprecedented, and the system may not be resilient to this level of pressure. This makes it very important to identify the thresholds of acceptable use, wherein the forest ecosystem would continue to be resilient to changes caused by human use.

Conservation Implications for Ecosystem Services

The results from this series of studies suggest that people are highly dependent on forest products for specific and non-specific needs. Their use and management of the forest may intentionally increase the relative abundance of some species, but may frequently and unintentionally decrease the species they value (for example, bamboo) as well as other species important for ecosystem functioning. Therefore, there appear to be some trade-offs in the system where direct species-specific ecosystem services are prioritised (*Madhuca indica* and *Diospyros melanoxylon*, etc.), while direct replaceable ecosystem services (wood, fodder) and indirect ecosystem services (hydrological flows, carbon sequestration) may be compromised. Thus, changes in species composition may alter the ability of forests to provide services whose benefit may be less tangible, such as hydrology, carbon storage and wildlife habitat (Costanza et al. 1997; Naeem et al. 1999). For example, conversion of tropical dry forests to scrub is associated with poor soil characteristics (Mehta et al. 2008), and different species composition in the same area is associated with lower infiltration and evapotranspiration (Krishnaswamy et al. 2013). Yet, people in the landscape only reported the loss of bamboo, suggesting that the local people have not noticed changes in species that they value less.

Most decisions regarding forest use and management occur at very local scales, and may even be made at the level of the individual; and there are differences between legal forest use activities prescribed by the Forest Department and on-ground forest use. We were able to obtain fine-scale variations in forest impact as we relied on more accurate forest use patterns available to us through trust and kinship networks instead of using broader scale management categories that are not respected on the ground. If local people are convinced of the value of *satkatha* species, communicating the loss of these species to local communities using on-ground networks of trust may lead local people to change their use of *satkatha* species and manage fire and grazing differently.

There is a strong impetus to conserve the 'natural' world for purposes of indirect ecosystem services such as wildlife conservation, carbon storage and hydrology, but the natural world is not a static entity. It consists of biophysical variations and disturbances, and there may be resilience (Allen et al. 2002) even in systems with large oscillations. In conserving ecosystems, we need to understand the dynamism in a system in order to understand whether conservation will work. If we conserve a static entity that we assume represents the natural system, there may not be enough resilience within that system to maintain ecological processes in future.

In India's context, given the long-term interaction of humans and the natural system, ecological dynamics is particularly important to understand. Paleontological and historical records have established the dynamic nature of ecosystems, have led to a greater emphasis on ecosystem resilience instead of ecosystem stasis (Allen et al. 2002; Willis and Birks 2006), and have shed light on many human practices previously considered degrading. For example, long-term remote sensing data in the Sahel was used to show that desertification there was climate-induced (Tucker et al. 1985) and not due to human practices such as overgrazing (Anyamba and Tucker 2005).

In central India, it appears as if fire-resistant species have been abundant in the landscape for a long time, with increasing relative abundance during dry climates, while no trend can be seen for cattle-resistant species. People are affecting species and diversity composition of forests through intentional and unintentional ways, and this is affecting ecosystem services. However, given the larger picture (obtained from palynological evidence), this may be a blip in geological time to which the system is resilient. While it is unclear whether the frequency and intensity of fire has increased to a point where forests can no longer provide services in future, human pressure may also have exceeded past thresholds. This is particularly in reference to cattle grazing that impacts relative abundance of species that are susceptible to herbivory such as *Miliusa tomentosa*, *Garuga pinnata* and *Shorea robusta*, which in turn provide food for cattle and wild animals.

While this series of studies illuminated patterns, we need mechanistic studies that examine changes in species composition and relative abundance caused by reduced forest use as well as long-term palynological studies. This will help us understand whether the forest is resilient to contemporary pressures or whether the forest ecosystem has crossed a tipping point and will not be able to provide the same ecosystem services in future.

Ultimately, management of the forest hinges on the value different stakeholders (managers, conservationists, local users, etc.) place on different species in the forest, and how they prioritise functions of the forest. Discussions between stakeholders may lead to prioritisation of different functions, but whatever our final goal, we still need to understand the biophysical variation in the forest to understand its resilience, which is critical in allowing our forests to function in the face of the changes in future.

References

Agarwala, M., R. Defries, Q. Qureshi, and Y. V. Jhala. 2016a. 'Factors Associated with Long-term Species Composition in Dry Tropical Forests of Central India', *Environmental Research Letters* 11(10): 1–19. Available at http://dx.doi.org/10.1088/1748-9326/11/10/105008.

———. 2016b. 'Changes in the Dry Tropical Forests in Central India with Human Use', *Regional Environmental Change* 16(1): 5–15.

Allen, C. D., M. Savage, D. A. Falk, et al. 2002. 'Ecological Restoration of Southwestern Ponderosa Pine Ecosystems: A Broad Perspective', *Ecological Applications* 12(5): 1418–33.

Anyamba, A., and C. J. Tucker. 2005. 'Analysis of Sahelian Vegetation Dynamics using NOAA- AVHRR NDVI Data from 1981–2003', *Journal of Arid Environments* 63(3): 596–614.

Black, A. E., P. Morgan, and P. F. Hessburg. 2003. 'Social and Biophysical Correlates of Change in Forest Landscapes of the Interior Columbia Basin, USA', *Ecological Applications* 13(1): 51–67.

Brandis, D. 2007. *Forest Flora of North-West and Central India: A Handbook of the Indigenous Trees and Shrubs of Those Countries.* Dehradun: Bishen Singh Mahendra Pal Singh.

Bowman, D. M. J. S., J. K. Balch, P. Artaxo, et al. 2009. 'Fire in the Earth System', *Science* 324(5926): 481–84.

———. 2011. 'The Human Dimension of Fire Regimes on Earth', *Journal of Biogeography* 38(12): 2223–36.

Burrows, N. D. 2008. 'Linking Fire Ecology and Fire Management in South-west Australian Forest Landscapes', *Forest Ecology and Management* 255(7): 2394–2406.

Census of India. 2011a. *Census Data Purchased for 2001 and 2011* (Ministry of Home Affairs, Government of India). Available at http://censusindia.gov.in.

———. 2011b. *ST Census 2011.* Available at http://saipdata.awardspace.com/madhyapradesh_district-wise_st_population.htm (11 February 2017).

Chauhan, M. S., and M. F. Quamar. 2010. 'Vegetation and Climate Change in Southeastern Madhya Pradesh during Late Holocene, Based on Pollen Evidence', *Geological Society of India* 76(2): 143–50.

———. 2012a. 'Mid-Holocene Vegetation vis-a-vis Climate Change in Southwestern Madhya Pradesh, India', *Current Science* 103(12): 1455–61.

———. 2012b. 'Pollen Records of Vegetation and Inferred Climate Change in South-western Madhya Pradesh during the Last ca. 3800 Years', *Geological Society of India* 80(4): 470–80.

Chauhan, M. S., K. Kumar, M. F. Quamar, and A. Sharma. 2013. 'Correlation of Data on Loss-On-Ignition and Palynology for Late Quaternary Climate Change in Southwestern Madhya Pradesh, India', *Current Science* 104(3): 299–302.

Costanza, Robert, Ralph d'Arge, Rudolf de Groot, et al. 1997. 'The Value of the World's Ecosystem Services and Natural Capital', *Nature* 387: 253.

Crook, C., and R. A. Clapp. 1998. 'Is Market-oriented Forest Conservation a Contradiction in Terms?', *Environmental Conservation* 25(2): 131–45.

DeFries, R., P. Mondal, D. Singh, et al. 2016. 'Synergies and Trade-offs for Sustainable Agriculture: Nutritional Yields and Climate-resilience for Cereal Crops in Central India', *Global Food Security* 11(special issue): 44–53.

Dutta, T., S. Sharma, B. H. McRae, et al. 2016. 'Connecting the Dots: Mapping Habitat Connectivity for Tigers in central India', *Regional Environmental Change* 16(supplement 1): 53–67.

Furley, Peter A., M. Robert Rees, Casey M. Ryan, and Gustavo Saiz. 2008. 'Savanna Burning and the Assessment of Long-term Fire Experiments with Particular Reference to Zimbabwe', *Progress in Physical Geography* 32(6): 611–34.

Gadgil, M., and R. Guha. 1993. *This Fissured Land: An Ecological History of India.* Berkeley: University of California Press.

Guha, S. 1999. *Environment and Ethnicity in India, 1200–1991.* New York: Cambridge University Press.

Heywood, V. H., and J. M. Iriondo. 2003. 'Plant Conservation: Old Problems, New Perspectives', *Biological Conservation* 113(3): 321–35.

Hiremath, A. J., and B. Sundaram. 2013. 'Invasive Plant Species in Indian Protected Areas: Conserving Biodiversity in Cultural Landscapes'. In L. C. Foxcroft, P. Pysek, D. M. Richardson, and P. Genovesi (eds), *Plant Invasions in Protected Areas: Patterns, Problems and Challenges,* pp. 241–66. The Netherlands: Springer.

Houghton, R. A. 2012. 'Carbon Emissions and the Drivers of Deforestation and Forest Degradation in the Tropics', *Current Opinion in Environmental Sustainability* 4(6): 597–603 Available at http://dx.doi.org/10.1016/j.cosust.2012.06.006.

Keane, R. E., James K. Agee, Peter Fulé, et al. 2008. 'Ecological Effects of Large Fires on US Landscapes: Benefit or Catastrophe?', *International Journal of Wildland Fire* 17(6): 696–712.

Keeley, J. E. 2006. 'Fire Management Impacts on Invasive Plants in the Western United States', *Conservation Biology* 20(2): 375–84.

Krishnaswamy, J., M. Bonell, B. Venkatesh, et al. 2013. 'The Groundwater Recharge Response and Hydrologic Services of Tropical Humid Forest Ecosystems to Use and Reforestation: Support for the "Infiltration-Evapotranspiration Trade-off Hypothesis"', *Journal of Hydrology* 498: 191–209.

Madhya Pradesh Forest Department. 2015. 'Profile of Development: Madhya Pradesh Forest Department'. Available at https://mpforest.gov.in/developmentscheme.html (11 March 2015).

Mehta, V. K., P. J. Sullivan, M. T. Walter, et al. 2008. 'Ecosystem Impacts of Disturbance in a Dry Tropical Forest in Southern India', *Ecohydrology* 1(2): 149–60.

Mondal, P., M. Jain, A. W. Robertson, et al. 2014. 'Winter Crop Sensitivity to Inter-Annual Climate Variability in Central India', *Climatic Change* 126(1–2): 61–76.

Morrison, K. D. 1994. 'Monitoring Regional Fire History Through Size-Specific Analysis of Microscopic Charcoal: The Last 600 Years in South India', *Journal of Archaeological Science* 21(5): 675–85.

Naeem, S., Chapin III, F. S., Robert Constanza, et al. 1999. 'Biodiversity and Ecosystem Functioning: Maintaining Natural Life Support Processes', *Issues in Ecology* 4: 1–15. Available at https://www.esa.org/esa/wp-content/uploads/2013/03/issue4.pdf.

Olander, L. P., H. K. Gibbs, M. Steininger, et al. 2008. 'Reference Scenarios for Deforestation and Forest Degradation in Support of REDD: A Review of Data and Methods', *Environmental Research Letters* 3(2): 025011.

Prasad, A. 2003. *Against Ecological Romanticism: Verrier Elwin and the Making of the Anti-Modern Tribal Identity*. New Delhi: Three Essays Collective.

Preece, N. D. 2013. 'Tangible Evidence of Historic Australian Indigenous Savanna Management', *Austral Ecology* 38: 241–50.

Pyne, S. J. 2001. 'The Fires This Time, and Next', *Science* 294(5544): 1005–06.

Quamar, M. F., and M. S. Chauhan. 2012. 'Late Quaternary Vegetation, Climate as Well as Lake-Level Changes and Human Occupation from Nitaya Area in Hoshangabad District, Southwestern Madhya Pradesh (India), Based on Pollen Evidence', *Quarterly International* 263 (special issue): 104–13.

———. 2014. 'Signals of Medieval Warm Period and Little Ice Age from Southwestern Madhya Pradesh (India): A Pollen-inferred Late-Holocene Vegetation and Climate Change', *Quarterly International* 325: 74–82.

———. 2015. 'Pollen-based Vegetation and Climate Change in Southwestern Madhya Pradesh (India) During the Last 3300 Years', *Journal of Palaeontological Society of India* 60(2): 47–55.

Rangarajan, M. 1996. *Fencing the Forest: Conservation and Ecological Change in India's Central Provinces 1860–1914*. New Delhi: Oxford University Press.

Ratnam, J., W. J. Bond, R. J. Fensham, et al. 2011. 'When is a "Forest" a Savanna, and Why does it Matter?', *Global Ecology and Biogeography* 20(5): 653–60.

Ratnam, J., K. W. Tomlinson, D. N. Rasquinha, et al. 2016. 'Savannahs of Asia: Antiquity, Biogeography, and an Uncertain Future', *Philosopical Transactions of the Royal Society B: Biological Sciences*, 371(1703): 20150305.

Rudel, Thomas K., Ruth Defries, P. Gregory Asner, et al. 2009. 'Changing Drivers of Deforestation and New Opportunities for Conservation', *Conservation Biology* 23(6): 1396–1405.

Ryan, Casey M., and M. Williams. 2011. 'How does Fire Intensity and Frequency Affect Miombo Woodland Tree Populations and Biomass?', *Ecological Applications* 21(1): 48–60.

Saha, S., and H. F. Howe. 2003. 'Species Composition and Fire in a Dry Deciduous Forest', *Ecology* 84(12): 3118–23.

Saigal, S. 2008. 'Non-Timber Forest Products and Forest Governance: Synthesis Report Based on Three State-Level Studies Carried Out During 2006–2007', Forest Governance Learning Group, India.

Sankaran, M., D. J. Augustine, and J. Ratnam. 2013. 'Native Ungulates of Diverse Body Sizes Collectively Regulate Long-Term Woody Plant Demography and Structure of a Semi-arid Savanna', *Journal of Ecology* 101(6): 1389–99.

Sankaran, M., J. Ratnam, and N. P. Hanan. 2004. 'Tree–Grass Coexistence in Savannas Revisited—Insights from an Examination of Assumptions and Mechanisms Invoked in Existing Models', *Ecology Letters* 7: 480–90.

Sankaran, M., N. P. Hanan, R. J. Scholes, et al. 2005. 'Determinants of Woody Cover in African Savannas', *Nature* 438: 846–49.

Scheffer, M., S. Carpenter, J. A. Foley, et al. 2001. 'Catastrophic Shifts in Ecosystems', *Nature* 413(6856): 591–96.

Sharma, S., T. Dutta, J. E. Maldonado, et al. 2013. 'Forest Corridors Maintain Historic Gene Flow in a Tiger Metapopulation in the Highlands of Central India', *Proceedings of the Royal Society B: Biological Sciences* 280(1757): 20131506.

Sivaramakrishnan, K. 1999. *Modern Forests: Statemaking and Environmental Change in Colonial Eastern India.* Stanford, CA: Stanford University Press.

Skaria, A. 1998. 'Being Jangli: The Politics of Wildness', *Studies in History* 14(2): 193–215.

Staver, A. C., S. Archibald, and S. A. Levin. 2011. 'The Global Extent and Determinants of Savanna and Forest as Alternative Biome States', *Science* 334(6053): 230–32.

Thekaekara, T. 2014. 'A Good Fire', *The Hindu Business Line,* 4 April 2014.

Troup, R. S. 1983. *Silviculture of Indian Trees,* Vols 1–6. Dehradun: Forest Research Institute.

Tucker, Compton J., R. G. John Townsend, and Thomas E. Goff. 1985. 'African Land-Cover Classification Using Satellite Data', *Science* 227(4685): 369–75.

Willis, K. J., and H. J. B. Birks. 2006. 'What is Natural? The Need for a Long-term Perspective in Biodiversity Conservation', *Science* 314(5803): 1261–65.

Yumnam, B., Y. V. Jhala, Q. Qureshi, et al. 2014. 'Prioritizing Tiger Conservation through Landscape Genetics and Habitat Linkages' *PLOS One* 9 e111207.

5

Water Under the Bridge

Wetland Use and Abuse in India

NEHA SINHA

In the winter of 2011, ecologist Sumit Dookia visited a place called Najafgarh drain. The name of the area itself is an oddity: the word 'drain' suggests sewage, rot, and an absence of fecundity. But what Dookia found surprised him. The Najafgarh 'drain' was an old waterbody, about 51 kilometres long, which flows through Delhi and Haryana. The area was teeming with birdlife, and mammals were spotted utilising it. The Najafgarh drain itself comes from something usually considered more respectable than a drain—a river, the Sahibi River. The Delhi government notes that the Najafgarh area has flooded several times, as the channel carries water from the Sahibi River. It also observes that improved drainage in Haryana has led to more water entering Delhi through this drain.[1]

As Dookia and his students started exploring the area, they noted hundreds of greater flamingos, migratory ruffs, flocks of black-winged stilts, and several other water birds. Immediately, a chord was struck: the area, fairly close to Guru Gobind Singh Indraprastha University, was ripe to be adopted as a research site and a source for exploring local avian biodiversity.

While the Najafgarh wetland is called a 'drain', its identity as a biodiverse site—which includes a wetland or *Jheel* (lake)—belies both its name and this very identity. This mirrors a curious truth that surrounds wetlands in India: although many wetlands survive in the country, they have not been notified, identified or named in public record as wetlands or in any similar terms. These habitats exist as 'drains', or are seen as land rather than 'wetland', or just as sources of water rather than ecological systems in their own right[2] (Sinha 2016).

When Dookia and his students started studying the Najafgarh drain and its *jheel*, the horizon was mostly clear of obtrusions. The area lay in an isolated, seemingly forgotten part of Delhi. No one really went there; it was neither a popular tourist spot nor the first choice for birdwatchers. But the area was also not to remain as it was. Haryana shoulders the wetland from one side, with Delhi on the other side. On the Haryana shoulder, Dookia and his team witnessed apartment complexes rising quickly. They recall watching the real estate turn parts of the swampy wetland basin into land in Haryana, with the foreboding that the same would soon happen on the Delhi side, which is close to the fast-developing Dwarka sub-city.

Dwarka was made specifically to solve the housing crunch in Delhi. Differentiated from other housing projects in the National Capital Region (NCR), Dwarka was planned as a mini-city within the Delhi metropolis. Despite being located in a somewhat remote corner of West Delhi, this sub-city's size, scale, and location within the capital have made it a real estate favourite. Within Dwarka, the L Zone is yet to be converted into housing. On property sites like MagicBricks and 99acres, Dwarka's L Zone is advertised as having the cheapest land. The land here is cheap because the area is yet to be fully colonised; this Zone is the shadow sub-city waiting to rise from its wetland and surrounding precincts. Property advertisements say that most of the L Zone flats and residences will be completed between 2019 and 2021, and buyers are encouraged to invest in the area. Despite an ongoing property slump, it is understood that this Zone will be under brick and mortar soon. But for ecologists, nature enthusiasts and birdwatchers, the Najafgarh drain and its catchment deserve protection from land-use conversion. Among these competing interests, whose needs would be privileged?

The larger question is an even more basic one: while this drain clearly carries water and hosts waterfowl, is it a 'wetland'?

Wetlands, Wastelands and *Nullahs*

The Delhi Development Authority (DDA) describes the Najafgarh drain in the following words: 'It [The Najafgarh drain] is one of the very old drains which emanates from Rajasthan/Haryana as a rivulet, where phased improvements have been carried out over the years. The capacity of the drain is about 3,000–8,000 cusec in different segments, with very gentle gradient' (DDA 2007).[3]

In the larger policy framework, India's Wetland Rules, 2017[4] define wetlands as such:

> 'Wetland' means an area of marsh, fen, peatland or water; whether natural or artificial, permanent or temporary, with water that is static or flowing, fresh, brackish or salt, including areas of marine water the depth of which at low tide does not exceed six meters, but does not include river channels, paddy fields, human-made water bodies/tanks specifically constructed for drinking water purposes and structures specifically constructed for aquaculture, salt production, recreation and irrigation purposes (Wetland Rules 2017).

While the definition appears comprehensive, it leaves out at least two forms of inland water—flowing rivers and drains such as the Najafgarh drain.

The storm-water drain, also descriptively called the *'ganda naala'* (lit. dirty drain) has had a chequered and unwelcome reception in city planning. For the 2010 Commonwealth Games, parts of the Barapullah drain were covered to create a road for delegates. The attitude towards the *ganda naala* has been to neglect its ecological history as well as its importance (Baviskar 2013).[5] The term '*ganda naala*' is often used to give directions in India—and appears to have been accepted as a feature in several cities—but there is little reflection on how the *ganda naala* came to be. In upmarket Defence Colony in South Delhi, the 'dirty' Barapullah drain was covered to create parking lots, and a park and seating area before the high-profile Commonwealth Games. For all purposes, it seems that the drain can only offer recreation when it does *not* exist. The Najafgarh drain has also been part of a cover-up: the Delhi government spent several crores in building walls around the drain in Wazirabad. This was done to 'prevent people from throwing trash and tyres into the drain'.[6] It is not entirely clear whether this was also to protect the city from the perceived eye-sore that is the drain, or to protect the drain itself from the city.

And this is a vicious cycle. Nearly every year the local press reports on the bickering among the authorities over desilting drains in

Delhi. Before monsoon, the drains are to be desilted and cleaned, but this is usually not done on time. As a result, drains get blocked, breed mosquitoes, and generally emanate an unwholesome look, encouraging people to throw more garbage into them. Historic storm water channels have become, for most purposes in Delhi, part of a present that the city does not seem to want. Yet a small group of conservationists have been trying to protect storm water drains in the city. For them, storm water drains are part of the city's ecological history and future. Many argue that the encroachment of and engineering interventions in storm water drains, exacerbate flooding, among other problems.[7] A recent order by the National Green Tribunal specified that storm water channels should not be covered.[8]

But this has not deterred the Delhi government from continuing planning the covering of a chunk of the Najafgarh drain for a road, just like on the Barapullah drain. A 26 kilometre road, covering Kakrola to Wazirabad, is planned on the drain.[9] A feasibility study commissioned by the Public Works Department (PWD)—for a major portion of the stretch—has already been carried out.

As mentioned earlier in this chapter, at least part of the problem comes from the fact that India's Wetland Rules clearly leave out flowing water channels; this includes drains and rivers.

By leaving out rivers and storm water channels, the Rules seem to hint that these water bodies need management, but not biological conservation. Within these two categorisations of waterbodies too, a clear hierarchy is evident. If rivers contain spectacular biodiversity and are repositories of our history and culture, then drains seem to be their smelly, poor, ragtag cousins. We have to ask ourselves—would we privilege storm water channels more if they were called rivers? And do rivers remain 'rivers' when decimated by pollution, or do they become drains in our understanding and lexicon?

There have been pleas for bringing rivers into the ambit of the Wetland Rules, and thus, provide them with institutional and biological conservation. In the case of drains, the issue is even more complex, and less high-profile. Delhi has historically had many storm water channels, which have slowly become a trundling treacle of sludge and sewage. Storing and carrying rain water is an environmental service, but as the city grows, this particular 'service' is hardly appreciated. The 2015 floods in Chennai, thought to be exacerbated because of the covering of the Buckingham canal, showed that such services can no longer be ignored.

Clearly, streams and drains like Najafgarh would not be protected under the Wetland Rules, even though certain portions (such as lakes and *jheels*) may fulfil the wetland 'criteria' under current legislation. In the case of drains like Najafgarh, streams that historically carried storm water or river water have been converted to sewage channels all over India. The conversion from 'stream' to 'drain' has been rapid, and now there are moves to cover these drains. Covering up a drain is like washing hands off the issue, and pretending that the waterbody does not exist.

Is there a better understanding of wetlands, their services and peculiarities in other government dossiers apart from the Wetland Rules?

The Conservation Matrix

The Ministry of Environment, Forest and Climate Change (MoEFCC) offers a description of a wetland in its 2007 document, 'Conservation of Wetlands in India—A Profile'.[10] As compared to the Wetland Rules, 2017, this offers more detailed criteria.

The description says: '... wetlands must have one or more of the following attributes. At least periodically the land supports predominantly hydrophytes. The substrate is predominantly undrained hydric soil. The substrate is non-soil and is saturated with water or covered by shallow water sometime during the growing season of each year.'

This definition emphasises three key attributes of wetlands: (1) hydrology which is the study of the degree of flooding or soil saturation, (2) wetland vegetation (hydrophytes), and (3) hydric soils. This definition specifies various attributes of wetlands. It does not contradict the definition offered by the Ramsar Convention, an international convention on wetlands,[11] but specifies the parameters to be used for identification.

The emphasis here is on the physical and biological attributes of a wetland, stressing that the coming together of land and water creates vegetation communities. This could also be true for man-made wetlands.

But this problematises the definition of wetlands as per the 2017 and earlier 2010 Wetland Rules. As discussed earlier, while many wetlands like the Najafgarh *Jheel* and parts of the drain fulfil the conditions of being a wetland, there is a conservation and policy blind spot in the categorisation of drains and channels as per the Rules.

If one were to look beyond the water aspect of the wetland, the Wetland Rules give importance to the area (or land) around the wetland. The Rules point out that the 'zone of direct influence' or 'catchment region' of the wetland is *part* of the wetland conservation and management efforts. Thus, a wetland not only includes the waterbody which seasonally or perennially contains water, but *also* includes the land around the wetland which is part of its 'zone of influence'. By this logic then, portions of the Najafgarh *Jheel* as well as its drainage area or basin ought to be conserved as wetlands. But the problem lies in the fine print: the Najafgarh Drain or *Jheel* can only receive conservation benefit if it is identified and notified as a wetland by the relevant authorities.

Further, India's newly minted Wetland (Conservation and Management) Rules, 2017 have completely excluded man-made wetlands from their ambit. This is despite the fact that man-made activities in and around ecosystems like wetlands can both create as well as destroy ecological habitat. And, these activities interact with both man-made and natural wetlands, and can sometimes alter them dramatically. On this issue, the Water Policy of India[12] says: 'Urban settlements, encroachments and any developmental activities in the *protected* upstream areas of reservoirs/water bodies, key aquifer recharge areas that pose a potential threat of contamination, pollution, reduced recharge and those that endanger wild and human life should be strictly regulated' (my emphasis).

At the same time, the Policy notes: 'Encroachments and diversion of water bodies (like rivers, lakes, tanks, ponds, etc.) and *drainage channels* (irrigated area as well as urban area drainage) must not be allowed, and wherever it has taken place, it should be restored to the extent feasible and maintained properly' (my emphasis).

Thus, the Policy seems to make two suggestions: first, that water bodies *ought* to be protected. This echoes India's Wetland Rules, which also say that wetlands of biological importance need protection. But when faced with the very real issue of encroachment and developmental challenges, the Policy suggests that such activities should be regulated in *protected* upstream areas. Thus, if we ought to protect wetlands, then the process for demarcation or identification becomes all-important.

The *National Wetland Atlas of India*[13] says that India has a total of 7,57,060 wetlands. It identifies 11,747 rivers and streams in this figure. A single reading of the Atlas therefore reveals that India has an abundance of wetlands. But as is the case of the Najafgarh Drain and *Jheel*, we have wetlands that ought to be conserved but are not actually notified or identified.

In terms of conservation, the Wetland Rules, 2010, suggest a set of priorities. These include 'Ramsar sites, wetlands that are ecologically sensitive and important, such as national parks, marine parks, sanctuaries, reserved forests, wildlife habitats, mangroves, corals, coral reefs, areas of outstanding natural beauty, historical or heritage areas or areas rich in genetic diversity'.

In the Wetland Rules, 2017, the definition of what could 'qualify' as a wetland shrinks further. The Rules state that wetlands do not include 'river channels, paddy fields, human-made water bodies/ tanks specifically constructed for drinking water purposes and structures specifically constructed for aquaculture, salt production, recreation and irrigation purposes'.

One way of looking at the issue is attempting to understand how terrestrial, land-based ecosystems are conserved.

For the conservation of terrestrial ecosystems, India has several protected areas (PAs). For the conservation of areas that fall outside the PA network, India has the National Biodiversity Targets, created under the framework provided under the Convention on Biological Diversity, to which India is a signatory. Target 6 calls for the conservation of a network of ecologically important sites, which include the Important Bird Areas (IBAs). IBAs are sites which hold threatened bird species, congregations of birds above 20,000 in number, or endemic and restricted range species.[14] It is significant to note that wetlands are an important component of IBA sites.[15]

The MoEF's National Wetland Conservation Programme (NWCP) suggests the following criteria for identifying wetlands: Glaciatic, Tectonic, Oxbow, Lagoon, Crater, Salt Water, Urban, Pond/Tank, Man-made, Reservoir, Mangrove, Coral Reef and Creek (including thermal springs) wetlands.

The IBA list in India, identified by the Bombay Natural History Society and Birdlife International, has recently been updated. More than 550 IBAs have been identified. Out of these, 225 contain wetlands. These sites typically hold a large number of migratory birds, or are visited by critically endangered or restricted range species (see Table 5.1).

Table 5.1: Analysis of Values Ascribed to Wetlands by Different Documents

Document Name	*Value Described*	*Derivative of Value*
Wetland Rules, 2017[16]	Wise Use of Wetlands	Wetlands that are natural deserve protection; man-made wetlands like tanks and salt pans do not need biological conservation.
Wetland Rules, 2010[17]	Ecological Sensitivity	Areas outside PAs, eco-tones, areas with unique or fragile ecosystems. Ecological functions.
Wetland Rules, 2010, 2017	Natural Beauty	Universal appeal, unique features, tourism value.
Wetland Rules, 2010, 2017	Historical or Heritage Value	Value awarded by man, may or may not be derived from ecological functions.
Wetland Rules	Genetic Diversity	Biological importance, taxonomic importance, diversity.
National Wetland Atlas[18]	Man-made Wetlands	Use of water by man.
National Wetland Atlas	Natural Wetlands	Natural occurrence of water body and waterlogging.
Important Bird Areas[19]	Congregations of Birds More than 20,000	Ecological importance of wetland or wetland complex on migratory flyway.
Important Bird Areas	Existence of Critically Endangered or Endemic Bird Species	Wetland as habitat and refuge for key species.
Ministry of Environment Water Day Document[20]	Biological and Hydro-Geological Values, Shape and Vegetation, Origin	Ecological functioning and biodiversity.

Source: An analysis of the wetland descriptions made by different legislations or documents such as the Wetland Rules, 2017, or the Wetland Rules, 2010.

Should these different prioritisation schemes throw up a greater diversity of identified wetlands? The results are not very encouraging. Let us look at the state of Tamil Nadu, which, as per the Wetland Atlas, has a large number of wetlands. The Atlas finds that Tamil Nadu has 9,02,534 hectares of land under wetlands. Out of these, 3,16,091 hectares are categorised as lakes and ponds. There are 56,419 hectares of reservoirs and barrages, and 1,36,878 hectares of rivers and streams, and 2,37,613 hectares of man-made tanks and ponds. This appears to be a comprehensive assessment for a state as large as Tamil Nadu.

The IBA list shows that there are 20 wetland IBAs in Tamil Nadu, which correspond to sites with high avian and biodiversity values. A recent affidavit submitted by Tamil Nadu in an ongoing case on wetland conservation said that it is actively conserving 23 wetlands. Out of these, the majority—13 wetlands—are PWD tanks, which are part of existing sanctuaries. These include the PWD tanks in Vedanthangal, Karikili, Pulicat, Vettangudi, Kanjirankulam, Chitrangudi, Udayamarthandpuram, Vaduvoor, Koonthankulam, Karaivetti, Vellode, Melaselvanur, Theerthangal, Sakkarakottai and Ossudu bird sanctuaries. The other wetlands are part of reserved forests, a Ramsar site (Pulicat) or other sanctuaries.

There are at least two takeaways here. First, PWD tanks—managed heavily by the human hand—are being conserved as wetlands. Second, there seems to be no specialised effort to search for or notify wetlands that are outside known classification schemes such as PAs or reserved forests. A list of 23 sites is a sore climb-down from the nearly 9,00,000 hectare area wetland figure the Wetland Atlas threw up. While each wetland will not fulfil all the criteria suggested in Table 5.1, it is clear that wetlands are not being protected adequately. Notably it seems wetlands that do not *appear* familiar, spectacular, or already part of protected or reserved forests, are being missed out.

Perhaps the manner in which wetlands have historically been treated gives a clue as to why they face so much official neglect. For many years, wetlands have been considered as wastelands. This is widely understood to be so because historically wetlands did not generate revenue, unlike land (Dutta, pers. comm.).[21]

The Story of the *Tampara Poramboke, Beela Gayalu*

In Andhra Pradesh, a thermal power project by Nagarjuna Construction Company (NCC) was planned on a wetland in Sompeta.[22] The story of what followed is shocking. Farmers and fisherfolk who used the wetland protested against the setting up of the project. Verbal dissent followed physical protest. This protest lasted many days; women, too, joined in. The police had an altercation with the locals. Tragically, the protesters were shot at. Two fishermen lost their lives,[23] and several were injured. A report by the Comptroller and Auditor General (CAG) for the state of Andhra Pradesh had some revealing findings.[24] First, the report said that the Sompeta wetlands were not revealed to be active wetlands.[25] Second, the area was actually portrayed as a 'wasteland' and as 'non-cultivable'. Thus, the Environmental Impact Assessment (EIA) for the project did not consider the ecological values provided by the wetland, but labelled it as unproductive *land*.

Environmentalist Mrutyamjaya Rao[26] has documented several bird species in Sompeta: 'You don't actually need evidence of the fact that the area is a wetland. One can physically see the water in the area. Yet the area has been described as barren and uncultivable' (Rao, pers. comm.).[27] Interestingly, the very name of the area in the local language connotes the land of the lotus. The area is referred to as *Tampara Poramboke* or *Beela Gayalu*. In Telugu, *tampara* means lotus or lotuses, and *beela* means waterbody.

The depiction of the area as uncultivable, infertile or a wasteland is an interesting one. Being uncultivable suggests that the area is barren; however in this case, the opposite is true. The area is uncultivable not because it is barren, but rather, because it is *waterlogged* for a large part of the year.

A petition filed in the National Green Tribunal,[28] against the environmental clearance given to the project quotes from an Andhra Pradesh government order, dated 15 September 2008:

> The District Collector, Srikakulam has also reported that the Tahsildar, Sompeta has submitted alienation proposals through RDO, Tekkai to an extent of Ac 1125.08 cts of Swamp lands locally called as Beela available in the above 4 villages of Sompeta Mandal and all the above lands are *Govt swamp land* which are situated in a low lying area, receiving the excess water from the

> Mukundasagaram Pydgam reservoir and the Mahindratanya Groin[29] and flood waters during rainy season from the villages lying in the western side …

It further states:

> … and due to receiving and stagnation of heavy water, it adversely affects surround paddy fields and agricultural crops by inundation during rainy season and *that the standing water stagnates throughout the year makes in accessible to the swamp except Mid May when water dried up. The said land are neither suitable for cultivation nor for commercial plantation except useful for inland Fisherman by catching fish for their livelihood to some extent.* (my emphasis)

In the same petition, the project proponent re-emphasises that the area is not fertile. Perhaps like the Najafgarh *Jheel*, which seems to have an identity crisis in the books of planners, the Sompeta wetlands, or the *Tampara Poramaboke* and *Beela Gayalu* are not useful unless they are annexed and their land use is converted.

The CAG indicts the government for not recognising the wetland in Chapter 7 of its report:

> Failure of the State Government in taking note of the environmental impact of this project resulted in alienation of the water body to NCC (Nagarjuna Construction Company), despite the latter not having obtained the requisite clearances for setting up the power plant from the Union Ministries of Power and Environment & Forests.

It also says:

> Special Chief Secretary & CCLA [Chief Commissioner of Land Acquisition] in his reply confirmed that the land in question was a swamp but stated that it was being used only by fishermen for catching fish for their livelihood and since it was not considered good enough for any other activity, it was allotted for setting up a thermal plant.

As per the above, a wetland can therefore be described as a 'low lying swampy land' but still not be given the ecological and actual identity of a wetland. Most states have notified either very large or prominent wetlands. India has 27 designated Ramsar sites. These are large, spectacular waterbodies. They fulfil the criteria mentioned in Table 5.1. At the same time, *unremarkable* wetlands, which will also fulfil the same criteria, are being left out of conservation and management efforts.

Ecological Issues

While the lack of recognition of the ecological identity of wetlands is now obvious, it also occurs *despite* the ecosystem services they provide.

In the aftermath of a large-scale flood in Chennai in 2015, a couple of popular theories surrounding the flood emerged. Some blamed construction on top of the Buckingham Canal for exacerbating the flood. The others said that drains should be desilted to avoid flooding in the future. The Chennai airport, which is built on the floodplains of the Adyar River, was also flooded. As part of flood-proofing the airport,[30] the airport authorities cleaned local drains and storm water channels. Yet, the de-recognition of ecological needs of wetlands and waterbodies continues on an institutional level. It is noted that new threats like climate change may exacerbate sudden and unseasonal rain, and flood protection services by wetlands and river drainage systems are more critical to urban security than ever before.[31]

South Asia Network on Dams, Rivers and People (SANDRP), a research and advocacy group on wetlands, notes a cycle of causation and consequence in wetland management:[32]

> The rivers in Chennai and the Buckingham Canal have little flow round the year since supply from their catchments is intercepted and stored in four reservoirs. Consequently the exposed river bed and floodplains have been encroached upon by housing colonies. But when it floods and the rivers swell across their banks, these buildings block the passage of water into the sea.[33]

Encroaching upon a wetland by creating roads or housing is going against the 'nature' and 'ecological character' of the wetland ecosystem. The identification of wetlands should provide scope to better manage the wetland on the lines of ecology. The absence of identification or notification of a wetland means that its management is done based on whim. Further, wetlands and their catchments are treated as if they are *not* wetlands at all. This could not be more evident than in the case of the Najafgarh *Jheel* and *naala*, to whose story we will now return.

Two things are clear. First, the service of the Najafgarh drain in carrying water and flood control is overlooked. Second, the capital wants to make the drain and wetland into land in the form of a road, or a tourism-centred business enterprise.

Simultaneously, the potential of the Najafgarh drain to be 'developed' into roadways or land seems imminent. However, the area in Dwarka's L Zone is also a biological entity containing its own ecological integrity. In 1979, Harkirat S. Sangha recorded several endangered bird species in the Najafgarh *Jheel*.[34] 'Despite these [general] unfavourable conditions', he notes, 'we observed the following water birds. Grey heron, sarus crane, dabchick, painted stork, white ibis, little stint, common sandpiper, paddy bird, large egret, curlew, little cormorant, coot, blackwinged stint, little egret, cattle egret, and redwattled lapwing.'[35] In more recent times, the Najafgarh *Jheel* has also had congregations of the bar-headed goose and greater flamingos, both species which conservationists work on. Not only that, the area has also emerged as an important heronry, hosting more than 200 nests of birds like ibis, herons, cormorants and egrets. The globally endangered skimmer, known to be sensitive to water pollution, was also sighted in the Najafgarh drain.[36] Since the last decade, the Najafgarh *Jheel* has become a popular birdwatching destination. This is evident as citizens and birdwatchers have created Facebook groups dedicated to the area, with names such as 'Habitat Conservation Awareness for Najafgarh *Jheel*', and 'Najafgarhjheel Bird Sanctuary'.

Both the Najafgarh *Jheel* and drain are important from the biodiversity point of view. Yet, very different plans are afoot for the landscape.

As evidenced by property speculation and the DDA's own admission, the DDA proposes to build up the area of Najafgarh's basin. The Zonal plan for L Zone by the DDA plan notes: 'The zone is the largest of 15 zones of Delhi and is predominantly rural in character. Najafgarh is the only census town. Most of the area of Zone 'L' forms part of the Najafgarh drainage basin.' It then suggests the following plans for the area:

> Najafgarh Drain has potential for development as a lake which is to be utilized for the recreational activities as well as water

> recharging area. The development of the lake shall be taken up in a comprehensive manner by Govt. of Delhi, MCD, DDA along with the Haryana Urban Development Authority as part of the area of the proposed lake is falling in Haryana. Tourist infrastructure would be permitted on the banks of lake/artificial islands. (Page 17, DDA plan)

Once again it is worth emphasising that the DDA thinks the Najafgarh lake is ostensibly not existing well as a lake, as it holds the potential for being developed 'as a lake which is to be utilised for recreational activities'. Second, an appreciation of the wetland's beauty and biodiversity is the reason birdwatchers visit the spot, and this is also likely to interest tourists. The fact that the *Jheel* and drain are visited by migratory birds like pelicans and flamingos is a testimony to its ecological values, the sort that bolster nature-based tourism.

However, the DDA-based model of tourism incorporates invasive ideas of recreation and water sports rather than environmentally sustainable tourism that enhances existing ecological character. At least three other examples of water sport-based tourism, managed with no heed for ecology, stand out as examples that have not worked in Delhi. North Delhi's Bhalaswa Lake, an old oxbow lake, has lost most of its water.[37] Sanjay Lake in East Delhi is now managed for tourism and water sports; however, the lake has instead become a cesspool for sewage.[38] Naini Lake in Model Town has been managed solely as a water-boating destination for years, but the ecological needs of the lake have not been managed or addressed. As a result, the lake is now 'ecologically dead'.[39] Some locals have been recently demanding that the lake be covered, as they fear that the dirty water may breed mosquitoes. There is a citizens' movement going on to save the lake from any conversion; these citizens, under the banner of 'Save Naini Lake', stress that it is the novelty of having a waterbody near their homes, as opposed to the more common gardens and parks, which makes the wetland worth saving. This does not, however, seem to be common sentiment. Mostly, the lure of grabbing the land under and around the wetland seems to have triumphed in the capital. As per the Delhi government's own admission, as many as 89 waterbodies have been turned to parks, temples, green belts or sports complexes in the city.

Even in using wetlands as recreational tourism areas, it is clear that it is not the identity or functionality of the *wetland* that seems to lure city planners. Instead, the wetland is an unexciting canvas, to be painted with the thrill of water sport (even if the water is dirty!) or the potential to become a park or 'green belt'.

Although called 'protected', the management of some wetlands is akin to the protection of terrestrial ecosystems like forests (in PAs). A recent judgment by the National Green Tribunal provides a subtext. In a case to protect wetlands in Bangalore, it noted that construction activity had taken place in Bangalore without environmental clearances. These constructions, the Tribunal noted, were around both wetlands as well as *rajakaluves*, or storm water drains. If the belittled and neglected storm water drains were to have a moment in the spotlight, this is possibly it.

The Tribunal stressed that the 'loss of wetland interconnectivity in Bangalore district is attributed to enormous increase in population ... and the reclamation of tanks for various developmental activities' and that suggested that a 'buffer' be left around wetlands as well as *rajakaluves.*[40] It also directed the setting up of a committee to look at the adverse effects of constructions on wetlands and *rajakaluves.* Subsequently, the municipal corporation has cleared debris and encroachment around three *rajakaluves.*

The concept of a 'buffer' around a wetland is interesting. As per the Wildlife Protection Act of, 1972, tiger reserves have buffers around them for the purpose of modulating human dependencies and reducing human–wildlife conflicts.

On an ecological level, the buffer is set to ensure that the forest or natural area does not have a hard edge with human habitation. The fact that the Tribunal asks for a buffer around wetlands suggests that there could be multiplicity of human use, as well as the provision for flood plains. The concept closest to the 'buffer' in the Wetland Rules, 2017 is that of the 'zone of influence'. Not only does the zone of influence of the wetland need to be free of human encroachment, but this also calls for a greater involvement of local communities or individuals. However, this concept is now under greater threat than before.

Towards Even More Damage?

In understanding how policy treats wetlands in India, we have seen that conservation rules have left out rivers and streams; that zones of influence of wetlands, and indeed wetlands themselves, are often unceremoniously annexed from identification of wetlands; and that most wetland conservation schemes tend to favour only familiar wetlands under existing PAs. The Wetland Rules, 2017 mandate that wetlands will only be protected if they are identified by the state. The 2010 Wetland Rules said the state is meant to submit 'brief documents' to identify and notify wetlands, and the Rules also called for a Central Wetland Regulatory Authority (CWRA). However, this Authority has only met once since 2012. Till 2016, not a single wetland has been notified by states, as per their communication in an ongoing case in the National Green Tribunal.[41] It was only after a rap by the Tribunal that the CWRA met in 2016. In 2016, the government of India also circulated a new draft for Wetland Rules. This draft received wide criticism from environmentalists as being a step down from the 2010 Rules.

Towards the end of 2017, the new Rules were notified. One of the concerns with the Wetland Rules, 2017 is the reduction of emphasis on wetland nature and what could really contribute to the ecological character of a wetland. The 2010 Rules placed heavy emphasis on 'ecological character' and the 'nature' of the wetland.

Significantly, the Wetland Rules, 2017 prescribed the criteria under which wetlands could be identified (see Table 5.1); this would help in circumscribing exactly what could give a wetland character. The Wetland Rules, 2010 gave the following indicative criteria for wetland identification:

> Wetlands recognised as Ramsar sites of international importance under the Ramsar convention as specified in the Schedule.
>
> Wetlands in areas that are ecologically sensitive and important, such as, national parks, marine parks, sanctuaries, reserved forests, wildlife habitats, mangroves, corals, coral reefs, areas of outstanding natural beauty or historical or heritage areas and the areas rich in genetic diversity,
>
> Wetlands recognised as or lying within a UNESCO World Heritage Site ...

Other than these, the following criteria were also cited:

> High altitude wetlands or high altitude wetland complexes at or above an elevation of two thousand five hundred metres with an area equal to or greater than five hectares,
>
> Wetlands or wetland complexes below an elevation of two

thousand five hundred metres with an area equal to or greater than five hundred hectares.

Any other wetland identified by the wetland authority and notified by the central government under the provisions of the Act.

This is what the Wetland Rules, 2017 say, on the matter of which wetlands ought to be protected:

(a) Wetlands categorised as 'wetlands of international importance' under the Ramsar Convention;

(b) Wetlands as notified by the Central Government, State Government and Union Territory Administration: Provided that these rules shall not apply to the wetlands falling in areas covered under the Indian Forest Act, 1927, the Wild Life (Protection) Act, 1972, the Forest (Conservation) Act, 1980, the State Forest Acts, and the Coastal Regulation Zone Notification, 2011 as amended from time to time.

In summation then, it is entirely up to committees to decide whether a wetland is a wetland or not. Under such a broad—or unspecific—mandate, it is entirely possible that several wetlands, especially those like the Najafgarh *Jheel*, would be left out of the identification and protection ambit. There is a second related issue that could go against maintaining ecological character. The 2010 Rules specified restrictions on certain activities in notified wetlands. This included 'activities within the zone of influence, as per definition of wetlands, which may directly affect the ecological character of the wetland'.

In the 2017 Rules, this portion finds no mention. It is as if restricting activities that affect ecological character are not of importance. And in terms of conservation and maintaining ecosystem services, management without restriction of ecologically harmful activities can be futile. This could also mean, in practical terms, that an important wetland would be managed for a value that has little to do with its ecology—like conserving a lake for running boats and tourism rather than biodiversity and ecology.

Let us go back to the case of the Najafgarh *Jheel*. A recent petition to conserve the area was filed by Indian National Trust for Art and Cultural Heritage (INTACH) under the National Green Tribunal. The Tribunal order dated 20 February 2017 states that the Haryana government has 'accepted Najafgarh *Jheel* as a waterbody'. The order further states that the Haryana government has decided to 'identify said water body' but the 'matter still has to receive approval from the competent authority in the State government'.

More than two years on, the Najafgarh *Jheel* has not been notified. Further, the Wetland Rules, 2017 were notified in September 2017. In another development, these rules have been challenged before the Supreme Court in 2018. The Court has directed the MoEF to file its response. The matter is currently under adjudication. So what then will happen to the Najafgarh *Jheel*—or drain—or wetland?

Wetlands do not look the same. Some wetlands look like land in dry seasons, some require brackish water (for example, the Pulicat lagoon on the eastern Coast), and some have partially submerged or swampy zones of influence. If the prescriptive 2010 Rules have not worked, do we have hope for wetlands being conserved under rules which are potentially far more accommodating for alternate land uses? Water remains a state and Central subject, but water also has the distinction of morphing into ecosystem and habitat, livelihood, and ecological identity. Ironically, while wetland ecosystems have a strong place and identity in human understanding, modern India also shows how easily it erases this identity, turning wetlands into land, roads, park and field. What hope would the students from Indraprastha University have for the Najafgarh *Jheel* and its local wetland, if its future is left solely to state administration, authorities who, under new rules, decide land use almost exclusively?

How India negotiates the distinctions of wetlands—rather than homogenising wetland as land, or protecting it only as forest—will set the tone for the future of wetlands in the country.

Notes

1. Irrigation and Flood Control Department, Govt. of Delhi, 2018, 'Flood Problem of Sahibi River', available at http://delhi.gov.in/wps/wcm/connect/DoIT_Irrigation/irrigation+and+flood+control/extra2/flood+problem+due+to+sahibi+river (accessed on 2 January 2018).
2. N. Sinha, 2016, 'Time Wetland Rules are Implemented by States', *Deccan Herald*, 10 February 2016, available at http://www.deccanherald.com/content/527979/time-wetland-rules-

implemented-states.html (accessed on 4 October 2016).

3. DDA, 2007, Draft Zonal Management Plan, Planning Zone L, West Delhi-III.
4. Ministry of Environment, Forests and Climate Change (MoEF), Govt. of India, G.S.R 1203(E) Wetlands (Conservation and Management) Rules, 2017, available at http://envfor.nic.in/content/gsr-1203e-wetlands-conservation-and-management-rules-2017 (accessed on 1 October 2017).
5. A. Baviskar, 2013, 'No Covering Up this Mess', *The Hindu*, available at http://www.thehindu.com/todays-paper/tp-opinion/no-covering- up-this-mess/article5246221.ece.
6. A. Shukla, 2006, 'Walls to be Constructed around Najafgarh Drain', *Hindustan Times*, 29 December 2006, available at http://www.hindustantimes.com/india/walls-to-be-constructed-around-najafgarh-drain/story-c3IBSLhnOBateULW1gvyGN.html (accessed on 5 October 2016).
7. South Asia Network on Dams, Rivers and People (SANDRP), 2016, 'Urban Wetlands and Floods', SANDRP blog, available at https://sandrp.wordpress.com/2016/04/05/urban-wetlands-and-floods/ (accessed on 20 October 2016).
8. Times News Network, 2014, 'Don't Cover Storm Water Drain: National Green Tribunal', *The Times of India*, 25 March 2014, available at http://timesofindia.indiatimes.com/city/delhi/Dont-cover-storm-water-drain-National-Green-Tribunal/articleshow/32623479.cms (accessed on 1 October 2016).
9. Times News Network, 2016, '10 Road Projects Back on Government Radar', *The Times of India*, 19 February 2016, available at http://timesofindia.indiatimes.com/city/delhi/10-road-projects-back-on-govt-radar/articleshow/51047049.cms (accessed on 4 December 2016).
10. Conservation Division, MoEF, Govt. of India, 2007, *Conservation of Wetlands in India: A Profile (Approach and Guidelines).*
11. Ramsar Convention, 1971, available at http://www.ramsar.org/about/history-of-the-ramsar-convention (accessed on 1 October 2016).
12. Ministry of Water Resources, Govt. of India, 2012, *National Water Policy.*
13. Space Applications Centre (SAC), Indian Space Research Organisation (ISRO), Ahmedabad, Govt. of India, 2013, *National Wetland Atlas: Wetlands of International Importance under Ramsar Convention.*
14. Birdlife International, 2017, 'Important Bird and Biodiversity Areas',

available at http://www.birdlife.org/worldwide/programme-additional-info/important-bird-and-biodiversity-areas-ibas (accessed on 1 October 2017).

15. M. Zafar-ul Islam, and A. R. Rahmani, 2004, *Important Bird Areas of India,* p. 32 (Oxford: Oxford University Press).
16. MoEF, Govt. of India, G.S.R 1203(E) Wetlands (Conservation and Management) Rules, 2017.
17. MoEF, Govt. of India, 'Draft Regulatory Framework for Wetlands Conservation for Comments', available at http://www.moef.nic.in/downloads/public-information/wetland-regulatory-framework.pdf (accessed on 1 October 2016).
18. SAC, ISRO, Ahmedabad, Govt. of India, *National Wetland Atlas.*
19. Islam and Rahmani, *Important Bird Areas.*
20. ENVIS Centre, Tamil Nadu, State of Environment and Related Issues, available at http://www.tnenvis.nic.in/Viewevents.aspx?Id=2261 (accessed on 2 October 2016).
21. Ritwick Dutta is an environmental lawyer practising in the National Green Tribunal.
22. B. V. Mahalakshmi, 2010, 'Nagarjuna Upbeat on Reviving Sompeta Power Plant Despite Green Hurdle', *The Financial Express,* 22 July 2010, available at http://www.financialexpress.com/archive/nagarjuna-upbeat-on-reviving-sompeta-power-plant-despite-green-hurdle/649486/ (accessed on 2 December 2016).
23. Mail Today Bureau, 2010, 'Fishermen Killed as Cops Open Fire in Srikakulam', *Mail Today,* 28 July 2010, available at http://indiatoday.intoday.in/story/fishermen-killed-as-cops-open-fire-in-srikakulam/1/106939.html (accessed on 17 October 2016).
24. Comptroller and Auditor General of India (CAG), 2012, *Report of 2012: Performance Audit on Civil Land of Government of Andhra Pradesh,* available at http://www.cag.gov.in/sites/default/files/audit_report_files/Andhra_Pradesh_civil_land_2012.pdf (accessed on 17 October 2016).
25. M. Suchitra, 2015, 'Sompeta Power Project: CAG Indicts Andhra Government for Allotting Wetland to Developer', *Down to Earth,* 4 July 2015, available at http://www.downtoearth.org.in/news/stay-on-land-allotment-for-thermal-power-plant-in-sompeta-33693 (accessed on 1 November 2016).
26. Mrutyumjaya Rao is an independent environmentalist and State Coordinator, Indian Bird Conservation Network, a network of people for conservation.
27. Mrutyumjaya Rao, pers. comm, Sompeta Wetlands.
28. National Green Tribunal, 2002, *T. Mohana Rao Versus Union of*

India, Written Submissions of Applicants.

29. Built to regulate water flow.
30. ET Bureau, 2016, 'Chennai Airport Completes its First Phase of Flood-Proofing after Deluge Last Year. Is it Enough?', *The Economic Times*, 13 November 2016, available at http://economictimes.indiatimes.com/industry/transportation/airlines-/-aviation/chennai-international-airport-completes-its-first-phase-of-flood-proofing-after-deluge-last-year/articleshow/55391191.cms?utm_source=contentofinterest&utm_medium=text&utm_campaign=cppst (accessed on 2 January 2017).
31. Krishna S. Vatsa, 2016, 'Be Warned, Urban Foods are Here to Stay', *The Wire*, 8 January 2016, available at https://thewire.in/18658/be-warned-the-urban-flood-is-here-to-stay/ (accessed on 2 January 2017).
32. SANDRP, 2016, 'Urban Wetlands and Floods'.
33. ibid.
34. H. S. Sangha, 1979, 'Najafgarh Jheel', *Newsletter for Birdwatchers* 19(12): 12.
35. ibid.
36. Vivek Menon and Tara Gandhi, 1992, 'Sighting of Skimmer at Najafgarh, Delhi', *Newsletter for Birdwatchers* 32(7–8): 17.
37. N. Sinha, 2009, 'Bhalaswa: Once There Was a Lake', *The Indian Express*, 11 August 2009, available at http://archive.indianexpress.com/news/bhalaswa-once-there-was-a-lake/500404/ (accessed on 20 October 2016).
38. Staff Reporter, 2016, 'The Story of a Disappearing Sanjay Lake', *The Hindu*, 13 May 2016, available at http://www.thehindu.com/todays-paper/tp-national/tp-newdelhi/the-story-of-a-disappearing-sanjay-lake/article5595544.ece (accessed on 2 January 2017).
39. Juhi Chaudhary, 2016, 'The Campaign to Save Delhi's Naini Lake', *thethirdpole.net: Understanding Asia's Water Crisis*, available at https://www.thethirdpole.net/2016/02/02/17446/ (accessed on 2 January 2017).
40. National Green Tribunal, 2015, Forward Foundation Versus Union of India, Original Application No. 222 of 2014, Judgment dated 7 May 2015.
41. Pushp Jain, 2016, 'Who is Responsible for Wetlands in India? Many and No One', *The Wire*, 3 September 2016, available at https://thewire.in/63605/wetlands-conservation-forests-cwra/ (accessed on 3 November 2016).

References

Baviskar, A. 2013, 'No Covering Up This Mess', *The Hindu*, available at http://www.thehindu.com/todays-paper/tp-opinion/no-covering-up-this-mess/article5246221.ece.

Birdlife International. 2017. 'Important Bird and Biodiversity Areas'. Available at http://www.birdlife.org/worldwide/programme-additional-info/important-bird-and-biodiversity-areas-ibas (accessed 1 October 2017).

CAG. 2012. *Report of 2012: Performance Audit on Civil Land of Government of Andhra Pradesh*. Available at http://www.cag.gov.in/sites/default/files/audit_report_files/Andhra_Pradesh_civil_land_2012.pdf (accessed 17 October 2016).

Chaudhary, Juhi. 2016. 'The Campaign to Save Delhi's Naini Lake', *thethirdpole.net: Understanding Asia's Water Crisis*. Available at https://www.thethirdpole.net/2016/02/02/17446/ (accessed 2 January 2017).

Conservation Division, MoEF, Govt. of India. 2007. *Conservation of Wetlands in India: A Profile (Approach and Guidelines)*.

DDA. 2007. Draft Zonal Management Plan, Planning Zone L, West Delhi-III.

ENVIS Centre, Tamil Nadu, State of Environment and Related Issues. Available at http://www.tnenvis.nic.in/Viewevents.aspx?Id=2261 (accessed 2 October 2016).

ET Bureau. 2016. 'Chennai Airport Completes its First Phase of Flood-Proofing after Deluge Last Year. Is it Enough?', *The Economic Times*, 13 November 2016. Available at http://economictimes.indiatimes.com/industry/transportation/airlines-/-aviation/chennai-international-airport-completes-its-first-phase-of-flood-proofing-after-deluge-last-year/articleshow/55391191.cms?utm_source=contentofinterest&utm_medium=text&utm_campaign=cppst (accessed 2 January 2017).

Irrigation and Flood Control Department, Govt. of Delhi. 2018. 'Flood Problem of Sahibi River'. Available at http://delhi.gov.in/wps/wcm/connect/DoIT_Irrigation/irrigation+and+flood+control/extra2/flood+problem+due+to+sahibi+river (accessed 2 January 2018).

Jain, Pushp. 2016. 'Who is Responsible for Wetlands in India? Many and No One', *The Wire*, 3 September 2016. Available at https://thewire.in/63605/wetlands-conservation-forests-cwra/ (accessed 3 November 2016).

Mahalakshmi, B. V. 2010. 'Nagarjuna Upbeat on Reviving Sompeta Power Plant Despite Green Hurdle', *The Financial Express*, 22 July 2010. Available at http://www.financialexpress.com/archive/nagarjuna-upbeat-on-reviving-sompeta-power-plant-despite-green-hurdle/649486/ (accessed 2 December 2016).

Mail Today Bureau. 2010. 'Fishermen Killed as Cops Open Fire in Srikakulam', *Mail Today*, 28 July 2010. Available at http://indiatoday.intoday.in/story/fishermen-killed-as-cops-open-fire-in-srikakulam/1/106939.html (accessed 17 October 2016).

Menon, Vivek, and Tara Gandhi. 1992. 'Sighting of Skimmer at Najafgarh, Delhi', *Newsletter for Birdwatchers* 32(7–8): 17.

Ministry of Water Resources, Govt. of India. 2012. *National Water Policy*.

MoEFCC, Govt. of India. 'Draft Regulatory Framework for Wetlands Conservation for Comments'. Available at http://www.moef.nic.in/downloads/public-information/wetland-regulatory-framework.pdf (accessed 1 October 2016).

———. G.S.R 1203(E) Wetlands (Conservation and Management) Rules, 2017. Available at http://envfor.nic.in/content/gsr-1203e-wetlands-conservation-and-management-rules-2017 (accessed 1 October 2017).

National Green Tribunal. 2002. *T. Mohana Rao Versus Union of India*, Written Submissions of Applicants.

———. 2015. Forward Foundation Versus Union of India, Original Application No. 222 of 2014, Judgment dated 7 May 2015.

Ramsar Convention. 1971. Available at http://www.ramsar.org/about/history-of-the-ramsar-convention (accessed 1 October 2016).

Rao, Mrutyumjaya, pers. comm, Sompeta Wetlands.

SAC, ISRO, Ahmedabad, Govt. of India. 2013. *National Wetland Atlas: Wetlands of International Importance under Ramsar Convention*.

SANDRP. 2016. 'Urban Wetlands and Floods', SANDRP blog. Available at https://sandrp.wordpress.com/2016/04/05/urban-wetlands-and-floods/ (accessed 20 October 2016).

Sangha, H. S. 1979. 'Najafgarh Jheel', *Newsletter for Birdwatchers* 19(12): 12.

Shukla, A. 2006. 'Walls to be Constructed around Najafgarh Drain', *Hindustan Times*, 29 December 2006. Available at http://www.hindustantimes.com/india/walls-to-be-constructed-around-najafgarh-drain/story-c3IBSLhnOBateULW1gvyGN.html (accessed 5 October 2016).

Sinha, N. 2009. 'Bhalaswa: Once There Was a Lake', *The Indian Express*, 11 August 2009. Available at http://archive.indianexpress.com/news/bhalaswa-once-there-was-a-lake/500404/ (accessed 20 October 2016).

———. 2016. 'Time Wetland Rules are Implemented by States', *Deccan Herald*, 10 February 2016. Available at http://www.deccanherald.com/content/527979/time-wetland-rules-implemented-states.html (accessed 4 October 2016).

Staff Reporter. 2016. 'The Story of a Disappearing Sanjay Lake', *The Hindu*, 13 May 2016. Available at http://www.thehindu.com/todays-paper/tp-national/tp-newdelhi/the-story-of-a-disappearing-sanjay-lake/article5595544.ece (accessed 2 January 2017).

Suchitra, M. 2015. 'Sompeta Power Project: CAG Indicts Andhra

Government for Allotting Wetland to Developer', *Down to Earth,* 4 July 2015. Available at http://www.downtoearth.org.in/news/stay-on-land-allotment-for-thermal-power-plant-in-sompeta-33693 (accessed 1 November 2016).

Times News Network. 2014. 'Don't Cover Storm Water Drain: National Green Tribunal', *The Times of India,* 25 March 2014. Available at http://timesofindia.indiatimes.com/city/delhi/Dont-cover-storm-water-drain-National-Green-Tribunal/articleshow/32623479.cms (accessed 1 October 2016).

———. 2016. '10 Road Projects Back on Government Radar', *The Times of India,* 19 February 2016. Available at http://timesofindia.indiatimes.com/city/delhi/10-road-projects-back-on-govt-radar/articleshow/51047049.cms (accessed 4 December 2016).

Vatsa, Krishna S. 2016. 'Be Warned, Urban Foods are Here to Stay', *The Wire,* 8 January 2016. Available at https://thewire.in/18658/be-warned-the-urban-flood-is-here-to-stay/ (accessed 2 January 2017).

Zafar-ul Islam, M., and A. R. Rahmani. 2004. *Important Bird Areas of India,* p. 32. Oxford: Oxford University Press.

6

Is Conservation Impossible?

The Case of Coastal Regulation in India

Kanchi Kohli and Manju Menon

Introduction: A Space of Contradiction

The coast is the cusp between the land and the sea. It is the zone of continued natural transition, of the creation of beaches and their breakdown—constantly in a state of flux. It is very visually appealing with its sand dunes, expansive beaches and rocky promontories jutting into the waters. A place of thriving plant and animal species diversity, coastal areas hold together huge colonies of corals, sea grass beds and tracts of mangrove forests that are the spawning and breeding grounds of a myriad number of species. This zone is also open to catastrophic natural events like tidal surges and cyclones. Coasts connect terrestrial, marine and human-built environments.

The 7,500-kilometre Indian coastline is considered to be one of the most biodiverse and unique ecosystems in the world, running across nine states, two Union Territories (UTs) and two island regions. Administratively it covers 69 coastal districts in mainland India; three in the Andaman and Nicobar Islands and one in the Lakshadweep Islands. According to the Centre for Coastal Zone Management and Coastal Shelter Belt, this area covers 43,230 sq. km of wetlands, it has 97 major river estuaries and 34 major lagoons. The area with 39 different species of mangroves goes across 6,740 sq. km, with 57 per cent being on the east coast, 23 per cent on the west coast, and 20 per cent in the Andaman and Nicobar Islands. There are 2,546 species of fish, 218 species of hard corals, 844 species of marine algae and 14 species of sea grasses.

When it comes to reptiles, the coast has five sea turtle and 26 sea snake species. Highly endangered marine mammals such as the Irrawaddy dolphin, the Ganges river dolphin and the sperm whale are all found on the coast (Centre for Coastal Zone Management and Coastal Shelter Belt 2016). There are also 31 Marine Protected Areas (MPAs) in India, covering an area of 6,271.2 sq. kms. The average size of these MPAs is 202.1 sq. km (Singh 2003).

According to the 2014 National Census, 14.2 per cent or 171 million people of India's population live in coastal districts. A large section of this population is dependent on a range of livelihoods including fishing, salt production, coastal agriculture and several other allied activities. While livelihood data for the coastal district is yet to be collated, the Central Marine Fisheries Research Institute (CMFRI) census of 2010 indicates that about four million or 8,64,550 families are directly engaged with fishing (Centre for Coastal Zone Management and Coastal Shelter Belt 2016).

Coastal lands are the gateway to very productive fishing grounds and are critical for fishing livelihoods, as the primary processing of the fish catch and the maintenance of craft and gear are carried out here. There are 1,94,490 fishing crafts found across India's coast, out of which 37 per cent are mechanised, 37 per cent are motorised and 26 per cent are non-motorised (Centre for Coastal Zone Management and Coastal Shelter Belt 2016). Coastal communities use these areas for numerous everyday activities. Boats are berthed on these beaches. Repairing of nets, drying of fish and fish selling are common sights here. Most importantly, the lands adjoining the sea front are critical for the housing of fishing communities. For these sea-faring families, reading the weather and the moods of the sea are critical to determining what their day at work will be like, far out at sea. A study shows that in Tamil Nadu, the average use of beach space for these activities is three times greater than the length of a coastal fishing village (Rodriguez et al. 2008). However, fisher families in most states have no land titles. Thus, they are vulnerable to being displaced as coastal land use policies are amended.

Vast stretches of these coastal lands have been leased to port trusts and harbours for trade. There are 13 major ports in India and several minor ports, with states like Gujarat and Maharashtra taking the lead with 40 and 53 ports each, respectively. Tamil Nadu, Kerala and Andhra Pradesh follow with 15, 13 and 12 ports each, respectively. Island areas on the coasts also have heavy port infrastructure: 23 in the Andaman and Nicobar Islands and 10 in Lakshadweep (Indian Port Association n.d.). Tourist facilities and private beach villas vie for this space due to their natural beauty. Large industries and power plants have been located on the coast since the 1960s. The coast is also the preferred location for thermal plants because of the availability of seawater that can be used for cooling and also since each power plant plans a captive jetty of its own to enable the direct supply of imported coal to fuel the power plant. A study found 27 existing coal-fired power plants on the coast with 59 more in the offing (Lakshmi et al. 2012). Besides these projects, coasts are preferred for the siting of special economic zones (SEZs), nuclear projects and other industries that require easy access to port facilities and the seafront.

The coast is therefore a space that epitomises policy contradictions. Should they be managed as conservation zones that support biodiversity-based livelihoods like fisheries? Or should they give way to the demands for housing, industry and infrastructure? Should they be treated as our first line of defence against disasters or should human use of the coast move further seawards? Without an understanding of these contradictions, much less a consensus on priorities and outcomes, India's coastal regulation jostles with these objectives on a regular basis to purportedly arrive at a 'balance of priorities' (Anon. 2011).

Coastal Zone Regulation in India

The coastal space is managed through a legal framework established by the Coastal Regulation Zone (CRZ) Notification, issued first in 1991 under the Environment (Protection) Act, 1986. There are four important phases that have shaped the current framework of coastal regulation in India. The first phase was marked by Prime Minister Indira Gandhi's letter to coastal states to take measures to protect coastal environments in 1981; second, the declaration of the CRZ through an executive notification in 1991; third, a series of policy reforms triggered after the 2004 tsunami; and fourth, the making and enforcement of the CRZ Notification, 2011.

Most analyses of the coastal regulation in India trace the history of this Notification to a letter written by Mrs Gandhi to governments of different coastal states in India on 7 November 1981 (Alvares 2011; Goenka 2000). Taking note of the unplanned development in coastal areas, her letter directed state governments to take measures to protect these spaces. The Environment Ministry, which was then called the Ministry of Environment and Forests (MoEF, under the Ministry of Agriculture), prepared a set of guidelines for protecting beaches in the country by the year 1997 (Mascarenhas 1999; MoEF 2005). However, the state and UT governments did not take these seriously as they did not have any legal backing (MoEF 2005).

In 1991, the MoEF issued the CRZ Notification. This is the first time that a coastal stretch of 500 metres on the landward side from the High Tide Line (HTL) was demarcated for regulated development. It separately laid down parameters for water bodies like river estuaries or creeks to be regulated based on the tidal influence on them. All of these together came to be understood as the coastal zone under the jurisdiction of the CRZ Notification. The Notification segregated the coastal land into three compartments, with CRZ 1 being the most ecologically fragile, covering nesting grounds of marine life, large tracts of mangroves and inter-tidal areas. CRZ 2 broadly encompassed urban spaces like coastal cities and towns, and CRZ 3 covered coastal zone in rural areas. While CRZ 1 to 3 have remained the same since the 1991 Notification, the 2006 Notification added CRZ 4, to cover territorial waters from the Low Tide Line (LTL) to 12 nautical miles out into the sea. Procedures laid out in the Notification regulate the setting up and expansion of any industry, operations and processes in the coastal zones (barring areas coming under MPAs, as those would be decided upon by bodies set up under the Wildlife Protection Act, 1972).

The CRZ Notification mandates prior approval for certain activities located in coastal zones. Construction of infrastructure, power generation, roads and tourism facilities require the project proponent to take permissions from the respective state-level Coastal Zone Management Authorities (CZMAs). The repair of houses, especially on the seaward side, requires people living within 500 metres from the HTL to approach the CZMA. The applicability of the CRZ Notification is flexible when it comes to defence projects. While land reclamation, bunding or disturbing the natural course of seawater is a prohibited activity in the CRZ, defence is one of the four kinds of activities upon which this prohibition does not apply. Defence-related ports and harbours are only allowed for defined strategic or defence purposes in high sea erosion areas. The Notification is not applicable for projects built in any of the CRZ areas prior to 1991, unless it relates to the expansion or addition of a new component to the same project.

Alongside the CRZ Notification, the Environmental Impact Assessment (EIA) Notification, 1994 regulates environmental decision-making on projects by establishing a process for gathering information about the potential impacts of projects, public participation in decision-making, and appraisal of the project by a team of experts. Whenever any activity proposed for siting in the CRZ is listed in the Schedule of the EIA Notification, 2006, permission needs to be sought from the Ministry of Environment Forests and Climate Change (MoEFCC). There is a special expert committee on Infrastructure and CRZ, which meets once a month and appraises applications for environment clearances in CRZ areas. In such cases, too, recommendations of the CZMAs form an important basis for taking decisions. Based on the content of the two regulations, it could be stated that the CRZ Notification is a 'siting' regulation, whereas the EIA Notification is a legal framework for environmental decision-making procedures. The CRZ Notification does not have procedures for the control of pollution as those are governed by pollution control laws. The Notification was amended 19 times in the first 17 years of its existence, with each amendment opening up the coast to more developmental activities that harm human, coastal and marine life (Menon et al. 2007).

Following the Indian Ocean tsunami in December 2004, the entire domain of official coastal planning, regulation and management began to be dominated by the scientific understanding of risk and resilience (Sridhar 2011). This also influenced the process of the review of the CRZ Notification that was being done by a committee headed by Dr M. S. Swaminathan, a renowned agricultural scientist. Instead of the much-needed reforms in overall coastal planning and governance, the Committee recommended a distancing of fisher and coastal communities from the coast. It suggested a Coastal Management Notification that recommended a 'setback' line for any activity (MoEF 2005). In its report, the tsunami overshadowed the 'non-events' of routine regulatory lapses that had already made coastal and fisher communities vulnerable to the disaster. The MoEF issued a draft CMZ Notification on 1 May 2008, amidst stiff opposition to the replacement of the 1991 CRZ law. In May 2009, when Jairam Ramesh took over as the Indian Minister of State (Independent Charge) at the MoEF, the issue of a new coastal regulation law was on the table again. The engagement of the National Fishworkers' Forum (NFF) and National Campaign for Protection of the Coast (NCPC)[1] with the MoEF, and the several official public hearings that were conducted to discuss this law and its implementation across the country led to the drafting and final issuance of the CRZ, 2011 (Anon. 2010; MoEF 2010a). The CRZ, 2011 is currently in operation and is a negotiated piece of regulation between the MoEF, scientists, industry representatives, fishing unions, International Financial Institutions (IFIs) like the World Bank, coastal protection networks as well as environmental groups (Centre for Environment Education [CEE] and MoEF 2010; Menon and Kohli 2010).

The CRZ Notification, 2011 is significant in many ways. It is the first time that a regulation has been put in place after several public hearings and engagement with a range of people who had a perspective on how coasts should be used, regulated and managed. Its Preamble says that the CRZ, which includes 500 metres to the landward side and 12 nautical miles into the sea, is put in place to:

> ensure livelihood security to the fisher communities and other local communities, living in the coastal areas, to conserve and protect coastal stretches, its unique environment and its marine area and to promote development through sustainable manner based on scientific principles taking into account the dangers of natural hazards in the coastal areas, sea level rise due to global warming …

There are three important elements of the CRZ, 2011. The first is the inclusion of the CZMA as the formal institutional mechanism to implement this executive-made law. The CZMAs were directed to be set up through orders of the Supreme Court in 1994, and have been operational since then, but the 1991 CRZ framework did not include this important change into the legal text. The second is the setting up of a formal timeline for the preparation of Coastal Zone Management Plans (CZMPs) and the demarcation of HTLs and LTLs. While there have been extensions to these timelines, the Notification has an intent to have this done, so that coastal planning can be less *ad hoc*.

The third inclusion has been most critical from the conservation point of view. It is the declaration of Critical Vulnerable Coastal Areas (CVCA), which are to be 'managed with the involvement of the local coastal communities including the fisherfolk'. Specific identified areas[2] were to be declared CVCAs through a consultative process with local fisherfolk and other coastal communities. This concept in the Notification brought attention to the need for the conservation of the coast, not as an exclusionary elitist system of protection, as has been demonstrated by the protected area (PA) system in India, but as a form of participatory management of coastal and marine resources. Integrated Management Plans are to be developed for these CVCAs.

Coastal Regulation Through Coastal Zone Management Authorities[3]

The CRZ Notification, 1991 had very broad mechanisms in place for the enforcement of the Notification. At that time, State Pollution Control Boards (SPCBs) were the only environmental institutions that could perform this function. A 1996 order of the Supreme Court dated 18 April 1996 recorded that SPCBs are both overworked and have a limited jurisdiction with regard to pollution. For ensuring the effective implementation of the CRZ and also the management plans that were to be put in place, the central government was asked to set up CZMAs. After this order, it took another three years for the CZMAs to fall into place, and since then, they remain the only institution in charge of overseeing the implementation of the CRZ Notification, at the state and national levels. Through their 17 years of existence, their performance has not been studied empirically and there has been no formal assessment of their role. Yet they were officially included in the 2011 Notification.

Data from a 2015 research study (Menon et al. 2015), undertaken to fill this gap in the assessment of environmental regulatory institutions, indicates that most State Coastal Zone Management Authorities (SCZMAs) do not operate independently. In almost all states, their offices are housed within the respective departments of environment. In fact, in most states the Chairperson is a representative from this department. In some instances, the SCZMAs find themselves in departments of Science, Technology and Environment. Interestingly, departments such as the SPCB and the state departments of forests and environment have always been part of the SCZMAs, as their representatives are usually the

Member Secretary and Chairman of the SCZMA. A study on the CZMAs (ibid.) found that scientists and scientific institutions do have a substantial representation in the authorities. In 2012, out of the 107 members of nine coastal states, over 22 were from academic institutions relating to sectors such as marine biology, oceanography, fisheries, botany, applied science and technology. Fourteen members from ministries and departments of pollution control, fisheries, and science and technology also belonged to the scientific fraternity.

The CZMAs perform a wide range of functions such as initiating conservation initiatives and awareness programmes, taking decisions on the use of the coastal zones, preparation of management plans, identification and action on violations of the law, as well as declaration of CVCAs.[4] What is interesting to also note is that the term 'conservation' is not used in any of the CZMA appointment orders that describe the functions of the SCZMA. However, tasks such as the identification of Ecologically Sensitive Areas (ESAs), areas highly vulnerable to erosion, and planning for the management of these areas have been listed to be undertaken by CZMAs from the time they first came to be set up in 1998. Since 2011, identification of economically important stretches has also been added as one of their functions. Members of the CZMAs are employed on a part-time basis to carry forward these tasks.

An analysis of the minutes of the National Coastal Zone Management Authority (NCZMA) and SCZMA meetings are one way to understand how these institutions are able to split their time amongst the various tasks required of them. At the national level, the CZMAs largely dealt with matters of reclassification of the coastal zones in the years 2013–14. Seventy-three out of the 157 agenda items dealt with requests to change the extent of regulation applicable in an area by deciphering whether it is in an ecologically fragile area, an urban zone or a rural one. In all these instances, the requests came only when regulatory provisions of the zone restricted proposed activities that could be undertaken there. SCZMA meetings held between 2010 and 2013 show that the institutions spend close to 46 per cent (in Goa) or 90 per cent (in Odisha) of their time appraising projects that seek CRZ approvals. Besides this, the SCZMA paid attention to the preparation of the CZMPs. States like Kerala and Goa spent as little as 1–2 per cent of their time finalising this document, which is supposed to be the basis of coastal planning under the CRZ (Menon et al. 2015). A substantial portion of the funds for SCZMA functioning comes from the fee they collect for project appraisals. This is called 'project application processing fee' or 'scrutiny fee' or 'coastal conservation fund', which is charged by different state coastal authorities for their work. Some SCZMAs have charged projects that seek approvals after the Environment Ministry made this suggestion to them through a letter dated 1 April 2000. The Goa SCZMA took a decision that it would be an 'unethical' practice to take a processing fee from project proponents, a rule that they kept for years till it was reversed in the 96th meeting of the SCZMA in December 2013. In states such as Odisha, once this practice was put in place, the state government stopped providing any financial support to the SCZMA.

Coastal Conservation Through The Coastal Regulation Zone?

The CRZ Notification, 2011 recommends the creation of special areas, CVCAs and ESAs, for '... protecting the critical coastal environment and difficulties faced by the local communities ...'. The Notification specifies that the CVCAs are to be declared through a process of consultation with local fisher and other communities inhabiting the area and depending on the coast for their livelihoods. Through these concepts, it expanded the need for coastal conservation beyond MPAs and CRZ 1 areas. However, the inclusion of these clauses brings the process of grant of approvals to coastal projects into a head-on collision with the declaration of CVCAs and ESAs. An intense conflict of priorities has been created by the inclusion of these provisions into a legal framework that has been operating mostly as an approval-granting mechanism. This is illustrated through two case studies of civil society efforts to get coastal areas demarcated and notified as CVCAs.

A Port in the Aghanashini Estuary

The Aghanashini is a river that originates in the Western Ghats of the Uttara Kannada (UK) district of Karnataka. It flows for approximately 140 kilometres through forested gorges and valleys, till it reaches the coastal stretch of the district, creating a wide estuary that finally joins the sea at the Aghanashini village in the

Kumta *taluk* (block) of the district. The estuary of the river is a flat expanse of water dotted with small islands and narrow creeks. Approximately 119.36 hectares of the estuary is covered with 12 different species of mangroves, and another 732.5 hectares is considered to have the potential for mangrove growth (Subhash Chandran et al. 2012).

A group of 1,800 bivalve collectors who are dependent on the estuary for a variety of estuarine shell fish, a local non-governmental organisation (NGO), Sahyadri Trust and the UK field office of Namati Environmental Justice Program, the Centre for Policy Research (CPR) submitted a proposal for the declaration of the Aghanashini River estuary as a CVCA to the Principal Secretary (Forests), Karnataka in June 2014. The proposal presented the biological importance of the estuary as well as its economic productivity. It records that hundreds of families in the villages around the estuary directly depend on it for their livelihoods through activities such as fishing, agriculture, collection of edible bivalves and crabs, shrimp aquaculture, traditional fish farming in the salt water rice fields (*gazni* lands), bivalve shell mining, salt production, sand removal, water transportation and other uses (Bivalve Collectors' Group et al. 2014; Boominathan et al. 2008).

The proposal presented an opportunity to recognise the importance of the Aghanashini estuary and also take planned measures to support both the conservation of the estuary and the livelihoods it supports. It quotes a study by the Indian Institute of Science (IISc), Bengaluru, which concluded:

> ... that annual revenue generated from Aghanashini estuary itself is about Rs 4.12 billion. Of these 94.64% comes from (3.9 billion) aquaculture in the estuarine belt. Fishing contributes about Rs 12.7 crores, followed by bivalve collection with a total of Rs 5.7 crore. Shell fish collection in the estuary contributes Rs 7.35 crores comprising bivalves, clams, oyster, mussels and other molluscs (Ramachandra et al. 2013).

The proposal was submitted twice to the Karnataka SCZMA, with copies to the Regional Director of the CZMA working at the district level (Bivalve Collectors' Group et al. 2014). However, the Karnataka SCZMA did not take any action on the CVCA proposal. Communication under the Right to Information (RTI) Act revealed that the CZMA Regional Director claimed not to have received this proposal from Uttara Kannada Regional office (Letter No. RDE/CRZ/KWR/RTI/2016-17/220, 20 July 2016). Meanwhile this area has been identified as a site for the construction of Karnataka's largest port. The project is proposed by the Karnataka State Industrial and Infrastructure Development Corporation (KSIIDC). The approval for this project was to be determined by the CZMA. The Karnataka SCZMA gave its nod for the construction of an all-weather port on 11 August 2015.[5] The CRZ law entrusts the conservation of areas like the Aghanashini estuary to the CZMA. However, the approval for the port took precedence and the CVCA proposal did not receive even a mention in this approval process. This is one example of how the CZMA balances priorities wherein one of the two competing values is made irrelevant or kept out of the decision-making process.

Since ports are listed in the EIA Notification, this port also requires an approval from the MoEF. A legally mandated environmental public hearing took place in March 2015 (Times News Network 2015). After two EIAs were done, the MoEFCC put forth several questions to the project proponent on the incompleteness and inadequacy of the EIA report prepared by the National Environmental Engineering Research Institute (NEERI),[6] a scientific institution of the Government of India (GoI). The port is to be built through one of the many public–private partnership models called DBOFT (Design, Build, Operate, Finance and Transfer). The public does not know the name of the private company who will build this port. The design is yet to be prepared by an identified concessionaire, and there exists no Detailed Project Report of the proposed port so far (Menon and Kohli 2015). A final decision on this proposal is pending as of 20 October 2016.

Closing in on the Kutch Coast

The Gulf of Kutch is one of the areas that the CRZ Notification, 2011 lists as a CVCA. The Kutch district in Gujarat has a coastline of 406 kilometres. The district is known for its 5-kilometre wide inter-tidal mudflats intercepted by numerous creeks, mangroves, seagrass beds and coral reefs. These mudflats exhibit a large species diversity of phytoplankton and zooplankton, which, together with mangroves, provide excellent feeding grounds for fish (Fishmarc and Kutch Nav Nirman Abhiyan 2010).

In April 2014, a proposal to declare a 38.6-kilometre long coastal stretch from the Luni fishing *bander* (harbour) to the Tuna fishing *bander* in the Anjar and Mundra *taluk* of Kutch as a CVCA was submitted to the Gujarat SCZMA. This stretch is the last remaining patch of free coastal space between the large gated estates of two of India's largest port facilities: one owned by Adani Port and SEZ Ltd. (APSEZ Ltd.), and the other belonging to the public sector Kandla Port Trust. Both projects have plans for expansion, although these are not justified based on existing project performance, demand for capacity addition or the larger context of port sector expansion being celebrated all along the Indian coast (Rodriguez and Sridhar 2010).

The area proposed for a CVCA comprises eight local fishing *banders* including Luni, Garudia, Bhadreshwar Randh, Kukadsar Bavadi, Veera, Dhavalvadi, Tuna and Takra. These *banders* are mostly classified as unsurveyed seashore or revenue wasteland, and are largely under common use by the fishing community. They practice artisanal fishing (with small boats) and *pagadia* (on foot) fishing that are well-suited to the wide inter-tidal ecology. Close to 800 families of the Wagher community are dependent on the fish catch at these *banders*. The fisheries economy dominated by the production of the dried and salted Bombay Duck (*Harpadon nehereus,* locally called *bombil)* is worth a total annual income of Rs 7.65 crores (1,572 tonnes) to Rs 17.66 crores across different fishing harbours.[7] This stretch of the coast also supports salt production in the inter-tidal area. The annual production of salt from these units goes beyond 5,50,000 tonnes (Namati, CPR 2014).

The CVCA proposal is also supported by the observations in the technical report on the 'Critical Habitat Information System of the Gulf of Kutch', prepared by the Integrated Coastal and Marine Area Management (ICMAM) Programme,[8] Ministry of Earth Sciences (MoES) in 2002. The MoES recommended for the 38-kilometre long coast between Mundra and west of Kandla creek to be kept free of developmental activities like ports. However, since the time the proposal was submitted, there has been no discussion on it or action by the Gujarat SCZMA. Instead, it has decided on proposals for the expansion of existing power plants and the setting up of new energy and infrastructure facilities. The Kandla Port expansion and establishment of integrated facilities has progressed, following approvals from both the Gujarat SCZMA and the Expert Appraisal

Committee (EAC) (MoEFCC 2014; 2016). These facilities are being developed through public–private partnership, i.e., between the public Kandla Port Trust and the Adani group that owns the APSEZ in the neighbouring part of the coast. The same group of companies has also proposed a power plant at the Bhadreshwar fishing harbour (Kutch Power Generation Ltd. 2014), and other energy companies in the region are also looking to expand their power generation facilities.

Even at the level of the MoEF, conservation priorities are rendered marginal. Two projects that have been given environmental approval by the MoEF are the Navi Mumbai International Airport and the Jaitapur Nuclear Power Plant, both on the Maharashtra coast. The website of the MoEF has uploaded long letters of justification given by Jairam Ramesh on why these projects have been approved despite strong opposition and impacts both on people and the environment (MoEF 2010a; 2010b; 2010c). The Jaitapur Nuclear Power Plant (6 x 1,650 Mega Watts [MW]) is located in Madban village of the Ratnagiri district of Maharashtra. The project was granted environmental clearance by the MoEF on 26 November 2010, just before the visit of the French President Nicolas Sarkozy to India in December 2010. The project was cleared with 35 approval conditions for the protection of the environment. The listing of the conditions came with a justification from the Union Minister of Environment and Forests, Jairam Ramesh, through a press release, which stated that the decision had been difficult and that a delicate balancing act had been done. There were weighty strategic and economic reasons to favour the project, even though the issues of marine biodiversity have been raised and also recognised by the MoEF (MoEF 2010b).

The Jaitapur approval was reported in the media, highlighting that it came barely a week after the approval granted to the controversial Navi Mumbai project with almost as many approval conditions. This clearance was regarded as being a 'satisfactory compromise', following MoEF's negotiations with the City and Industrial Development Corporation of Maharashtra Ltd. and the Ministry of Civil Aviation (Kohli 2011). The note of the then Minister, Jairam Ramesh, on the decision was made public. It stated that the project was of national importance and was approved based on optimisation of boundaries and additional environmental safeguards (MoEF 2010a).

Critical Vulnerable Coastal Areas, Scientifically Speaking

The above case studies are illustrative of the denial of coastal conservation in individual project-based decision-making by CZMAs or in creating space for conservation areas in industrialised or built-up coasts. This is the case even though the 2011 Notification has clearly stated clauses on conservation. This is not a denial only at the level of the SCZMAs. The CRZ Notification does not define critical vulnerability specifically. However, it directs that the MoEF develop guidelines on how this is to be done in consultation both with the state governments as well as coastal communities. Despite the SCZMAs seeking guidelines on how CVCAs are to be planned, they have not received a response from the MoEF.[9]

Instead, what has been taken up as a first step is the task of delineating ESAs in CRZ 1, which is being coordinated by scientists of the National Centre for Sustainable Coastal Management (NCSCM), constituted by the MoEF in 2012. The NCSCM is developing the criteria for the primary screening of ESAs listed in the 2011 Notification and a techno-scientific process to delineate ESAs. This comes at a time when the state and central governments have had serious conflicts over the very idea of ESAs, be it around PAs or in landscapes such as the Western Ghats (Gadgil 2014; Kapoor and Kohli 2012; Pinjarkar 2013). A process of Geographic Information System (GIS) mapping of these areas, building criteria for assessing the conservation value of ESAs, are all underway, and it is only after these steps that their management plans will be created.

For CVCAs, the NCSCM has a note titled 'Delineation and Mapping of Ecologically Sensitive Areas and Critical Vulnerable Coastal Areas along the Coast of India'. According to this note, guidelines for the demarcation of CVCAs and preparation of Integrated Management Plans will be prepared through consultations with ecologists and sociologists. The maps created through this process will have the boundaries of 'highly sensitive areas, sensitive areas and CVCAs'. Until these processes are completed, there is no hope that critical coastal habitats and livelihoods will be protected by the CRZ Notification. There is also no sign that these processes will learn from the experience of managing coastal and marine PAs, from the impacts of industrial projects, pollution and 'development' on the coast, or from the experiences of coastal communities who have lived with these impacts. These time-consuming and bureaucratic processes of creating definitions and scientific assessments have meant that no CVCAs or ESAs have been legally recognised as of October 2016, over six-and-a-half years after the CRZ Notification, 2011 mandated it to ensure the conservation of India's coasts.

The Volatility of Coastal Regulation

As stated earlier, the CRZ regulation went through 17 executive amendments before the 2011 law was brought into place. Each amendment was either in response to a demand made by a proponent who would benefit from the change in the law, or for the extension of deadlines for the preparation of CZMPs and the demarcation of tidal lines. Each change brought the CRZ closer to being unimplementable as a coastal conservation framework and instead allowed for large-scale commercial takeover of the coasts. But the Notification continued to carry an acknowledgement of the interconnectedness of the sea, the coast and the fishworker. This was further strengthened in the 2011 version of the CRZ, mainly due to the mobilisation efforts and advocacy by the independent trade union of fishworkers and the NCPC. As it is a Notification that is focused on a specific region such as the coast, representatives from coastal communities and union leaders were able to convince Jairam Ramesh of the need to make CRZ a tool to protect local coastal livelihoods. However, this solidarity between coastal communities is a fractured one in local areas where infrastructure projects are proposed. Coastal and fisher communities, like the rest of Indian society, are divided by caste, religion and livelihoods and therefore unable to create a consensual demand for development that is best designed to meet their long-term needs and well-being. At Tadri in Uttara Kannada, while the bivalve collectors are very concerned about the impacts of the port, the marine fishers who consider themselves a more privileged caste group have been less willing to join the collectors' opposition.

Since the Bharatiya Janata Party (BJP) government came to power, this legal framework put in place by the earlier government is yet again under review. In June 2014, the MoEF set up a six-member Committee, under the chairmanship of Dr Shailesh Nayak, Secretary, MoES. The purpose of the Committee was to examine issues of coastal states and also 'the errors or inconsistencies and procedural simplification in the said Notification'. The Committee members had extensive meetings with the governments of Maharashtra, Karnataka and Kerala, and also invited representatives of state and UT governments for a meeting in New Delhi. No consultations were carried out with coastal communities, fishing unions or private parties as per the information disclosed in the report. Even though the Committee submitted its report in January 2015, the MoEF did not make this report public until June 2016, and only after there was a direction from the Central Information Commission (CIC). The CIC had directed that policy processes such as the reforms to the CRZ regulation need to be publicly debated and information about the processes cannot be denied.

Once the contents of the report came to be publicly available, it presented two aspects of the continued volatility in the law-making for the coast. First, the Committee had recommended that the creation of a new draft CRZ Notification be issued, and second, the MoEFCC has relied on several recommendations of the Committee to shape amendments to the existing Notification even prior to the final submission of the report. Some of these amendments were to accommodate the state governments who wanted to build coastal highways, expand tourism facilities and promote real estate projects in regulated zones where they were otherwise not permitted. Analysis of the Committee's report and the amendments lead us to the conclusion that seven out of the eight amendments finalised from January 2015 to June 2016 were related to allowing more constructions in environmentally sensitive coastal zones (Khapre 2015; Namati, CPR 2016; Nandkumar 2016).

In December 2015, the environment ministry gazetted an amendment to the CRZ, which allowed reclamation for roads in CRZ areas. Prior to this amendment, reclamation was permitted only for roads on stilts as a protective measure for inter-tidal areas, mangroves or fishing spaces. The Maharashtra government particularly benefitted from this amendment, as it proposed to build the Mumbai coastal road project. Another amendment was to allow for the construction of monuments and memorials in the seaward side of the coastal zone, regarded as CRZ IV. This amendment came in February 2015 soon after the Shailesh Nayak committee finalised and submitted its recommendations to the MoEFCC. Due to this amendment, the Maharashtra government constructed a memorial statue of Chhatrapati Shivaji in the CRZ IV (Ghanekar 2015; Khapre 2015).

The 110-page Nayak Committee Report also proposes a 'Conservation, Protection and Management Framework for CRZ 1 areas' that are considered to be ecologically fragile areas and with dense mangroves, coral reefs and nesting grounds. It recommends the declaration of ESAs across coastal areas. But this recommendation makes no mention of the already ongoing efforts by agencies like the NCSCZM or the federal tussle on ESAs.

Is Conservation Impossible?

The implementation of the CRZ regulation has prioritised tasks related to processing and granting approvals rather than take forward urgent initiatives that would enable the conservation of India's coastline and support coastal livelihoods. Even though conservation is a key part of the CZMAs' mandate, only two SCZMAs addressed conservation, and that too, for less than 5 per cent of the total time period of the Authority's meetings. In almost all the coastal states, the majority of time in meetings is focused on project approvals (Menon et al. 2015). The MoEF, state-level authorities and the project proponents have worked towards clarifying procedures that would assist in reducing the time and work required to appraise applications for approvals and carry out necessary paperwork for the same. Even as projects are granted CRZ clearances as a routine matter, it is the efforts for conservation that have got stuck in bureaucratic procedures of defining frameworks, definitions and criteria on the one hand, and lack of prioritisation, on the other.

Coastal conservation efforts have been far more inclusive than terrestrial conservation in the Indian setting. Land-based conservation, with its emphasis on megafauna such as the tiger and the elephant, and a largely guns-and-guards approach dominated conservation practice in India. During this time, coastal conservation progressed through far more decentralised and community-level approaches. The CRZ Notification and CZMAs would do well to take lessons from these approaches to realise its objective of coastal conservation rather than wait for a centralised, single 'scientific' and formulaic model to achieve this. Since the 2011 Notification also includes the sea for the first time, the CRZ offers a unique opportunity to demonstrate participatory models of marine conservation along India's coasts.

The lasting value of the CRZ Notification and the agenda of coastal conservation, albeit with many amendments, dilutions and poor implementation, is the result of the vision and joint efforts of political leaders, fisher union leaders, coastal community representatives, as well as coastal conservationists and researchers. This solidarity is needed to realise coastal conservation and develop implementable plans that are inclusive, livelihood-friendly and robust enough to withstand the ever-increasing takeover of the coastal commons by private infrastructure projects.

Notes

1. The NCPC is a coalition of fishing unions, researchers, development agencies and marine conservationists that engaged regularly with the MoEF on the changes that were needed to the CRZ law and on reasons for the inadequacy of the CMZ.
2. The entire Sunderbans mangrove area and other identified ecologically important areas such as the Gulf of Khambat and the Gulf of Kutch in Gujarat, Malvan, Achara-Ratnagiri in Maharashtra, Karwar and Coondapur in Karnataka, Vembanad in Kerala, Gulf of Mannar in Tamil Nadu, Bhitarkanika in Odisha, Coringa, East Godavari and Krishna in Andhra Pradesh.
3. This section draws on the research study, M. Menon, M. Kapoor, P. Venkatram, et al., 2015, 'CZMAs and Coastal Environments: Two Decades of Regulating Land Use Change on India's Coastline. India' (New Delhi: Namati Environmental Justice Program, CPR).
4. Matters were clubbed under five categories: (i) Procedural: procedures, guidelines and administrative issues; (ii) CZMP: CZMP, zonation and reclassification; (iii) Project appraisals; (iv) Violations; (v) Conservation: conservation, ESAs, CVCAs, mangrove plantation, forestry, awareness building, education and training.
5. As recorded in the Minutes of first Meeting of the EAC (Infrastructure 2) held on 21–22 December 2015 at the MoEFCC, New Delhi.
6. ibid.
7. This data is based on the income of the Kutch Seafood Producer Company from 2000–05. It was also included in the proposal sent to the Gujarat Coastal Zone Management Authority (GCZMA) for the declaration of the Luni to Tuna CVCA on 18 April 2014.

8. The Integrated Coastal and Marine Area Management Project was established in 1998 and is supported by the MoES, Government of India. Its Directorate was established for implementing the International Development Association (IDA), World Bank-assisted Environmental Management Capacity Building Project.
9. In October 2012, the Maharashtra SCZMA decided to request the MoEF for guidelines regarding the preparation of management plans for Achara-Ratnagiri and Malvan-Sindhudurg. Till August 2013, no reply was received from the MoEF.

References

Alvares, C. 2011. 'Towards Ruin: The Coast is Finally Clear', *Outlook Magazine*, 24 January 2011.

———. 2010. 'Minister Attends Public Consultation on CRZ', *The Hindu*, 17 January 2010.

———. 2011. 'The Key Word is Balance' (Outlook Interviews—Jairam Ramesh)', *Outlook Magazine*, 7 February 2011.

Bivalve Collectors' Group, Sahyadri Trust and Namati Environmental Justice Program, CPR. 2014. 'Proposal for Declaration of Aghanashini Estuary as Critically Vulnerable Coastal Area (CVCA) under CRZ Notification, 2011', Submitted to Principal Secretary (Forests), Karnataka, June 2014.

Boominathan, M. Subash Chandran, M. D. Ramachandra, et al. 2008. 'Economic Valuation of Bivalves in the Aghanashini Estuary, West Coast, Karnataka', *ENVIS Technical Report*, 30 November 2008. Available at http://wgbis.ces.iisc.ernet.in/energy/water/paper/ETR30/index.htm (accessed 14 October 2016).

CEE and MoEF. 2010. 'Report of the Public Consultation with Fisherfolk and Community to Strengthen Coastal Regulation Zone (CRZ) Notification, 1991', New Delhi.

Centre for Coastal Zone Management and Coastal Shelter Belt. 2016. 'Database on Coastal States of India', 17 March 2016. Available at http://iomenvis.nic.in/index2.aspx?slid=758&sublinkid=119&langid=1&mid=1 (accessed 20 October 2016).

Fishmarc and Kutch Nav Nirman Abhiyan. 2010. 'Kutch Coast—People, Environment & Livelihoods'. Available at http://www.ceeindia.org/cee/pdf_files/kutch_ coast_study.pdf (accessed 30 May 2016).

Gadgil, M. 2014. 'Western Ghats Expert Ecology Panel: A Play in Five Acts', *Economic and Political Weekly* 49(18): 38–50.

Ghanekar, N. 2015. 'Centre will Tweak CRZ Notification to Pave the Way for Chhatrapati Shivaji Statue', *DNA News and Analysis*, 22 January 2015.

Goenka, D. 2000. 'The Fragile Coastline', *Seminar* (492), August 2000.

Indian Port Association Website. n.d. Available at http://www.ipa.nic.in/ (accessed 22 February 2017).

Kapoor, M., and K. Kohli. 2012. 'The Second Act: Centre–State Conversations on Ecologically Sensitive Areas (2009–2012)'. New Delhi: Kalpavriksh; Anand: The Duleep Matthai Nature Conservation Trust.

Khapre, S. 2015. 'Centre Notifies Draft on CRZ Amendments: Mumbai Coastal Road Project Can Now Begin', *The Times of India*, 7 July 2015.

Kohli, K. 2011. 'MoEF's Conditions Mean Nothing', *Civil Society* 8(4), February 2011.

Kutch Power Generation Ltd. 2014. 'Seeking Issuance of Terms of Reference for Proposed 3300 (5 × 660) MW Power Project of Kutchh Power Generation Ltd. at Village Bhadreshwar, Taluka Mundra in Kutchh District, Gujarat', 11 April 2014.

Lakshmi, Ahana, Auroilio Schiavina, Probir Banerjee, et al. 2012. 'The Challenged Coast of India', A Report Prepared by the Pondy Citizen's Action Network (PondyCAN), in collaboration with the Bombay Natural History Society (BNHS) and the Tata Institute of Social Sciences (TISS), October 2012.

Mascarenhas, A. 1999. 'The Coastal Regulation Zone of Goa: Oceanographic, Environmental and Societal Perspectives', *Current Science* 77(12): 1598–1605.

Menon, M., and K. Kohli. 2010. 'The World Bank and Environmental Policy Reform'. In Michelle Kelly and Deepika D'Souza (eds), *The World Bank in India: Undermining Sovereignty, Distorting Development*, 413–31. New Delhi: Orient BlackSwan.

———. 2015. 'A Strange Proposal to Build Karnataka's Largest Port Raises Serious (and Awkward) Questions', *Scroll.in*, 22 March 2015. Available at https://scroll.in/article/715364/a-strange-proposal-to-build-karnatakas-largest-port-raises-serious-and-awkward-questions (accessed 31 July 2015).

Menon, M., M. Kapoor, P. Venkatram, et al. 2015. 'CZMAs and Coastal Environments: Two Decades of Regulating Land Use Change on India's Coastline. India'. New Delhi: Namati Environmental Justice Program, CPR.

Menon, M., S. Rodriguez, and A. Sridhar. 2007. 'Coastal Zone Management Notification '07—Better or Bitter Fare?', *Economic and Political Weekly* XLII(38): 3838–40.

MoEF. 2005. 'Report of the Committee to Review the Coastal Regulation Zone Notification, 2005 (Chaired by MS Swaminathan)'. New Delhi: MoEF, Government of India.

———. 2010a. 'Report of the Public Consultation with Fisherfolk and Community to Strengthen Coastal Regulation Zone (CRZ)

Notification, 1991'. New Delhi: Centre for Environment Education, Ahmedabad; New Delhi: MoEF, Government of India.

———. 2010b. 'Press Release on Navi Mumbai International Airport', 22 November 2010.

———. 2010c. 'Press Release on NPCIL Jaitapur Power Park', 28 November 2010.

MoEFCC. 2014. 'Agenda for 136th Meeting of Expert Appraisal Committee for Projects Related to Infrastructure Development, Coastal Regulation Zone, Building/Construction and Miscellaneous Projects, 30 July, 2014–1 August, 2014', New Delhi.

———. 2015. Minutes of first meeting of the EAC (Infrastructure 2) held on 21–22 December 2015 at the MoEFCC, New Delhi.

———. 2016. 'Minutes of 7th Expert Appraisal Committee (Infra-2) Meeting Held on 29th June, 2016', New Delhi.

Namati Environmental Justice Program, CPR. 2014. 'Proposal on Luni to Tuna (Kutch) Critically Vulnerable Coastal Area (CVCA)', Submitted to the Gujarat Coastal Zone Management Authority on 18 April 2014.

———. 2016. 'Analysing the Implications of the Coastal Regulation Zone (CRZ) Committee Report'. Available at http://www.cprindia.org/news/5372 (accessed 24 February 2017).

Nandkumar, T. 2016. 'CRZ Amendment Raises Hackles', *The Hindu*, 11 January 2016.

Pinjarkar, V. 2013. 'States Get More Time to Declare Eco-Sensitive Zones', *The Times of India*, 29 May 2013.

Ramachandra, T. V., M. D. Subhash Chandran, N. V. Joshi, et al. 2013. 'Valuation of Estuarine Ecosystem of Uttara Kannada District, Karnataka', *Envis Technical Report 45*. Available at htttp://ces.iisc.ernet.in/biodiversity (accessed 24 February 2017).

Rodriguez, S., and A. Sridhar. 2010. 'Harbouring Trouble: The Social and Environmental Upshot of Port Growth in India', p. 62. Bengaluru: Dakshin Foundation.

Rodriguez, S., G. Balasubramanian, M. P. Shiny, et al. 2008. 'Beyond the Tsunami: Community Perceptions of Resources, Policy and Development, Post-Tsunami Interventions and Community Institutions in Tamil Nadu, India', p. 78. Chennai: UNDP/UNTRS; Bengaluru: ATREE.

Singh, H. S. 2003. 'Marine Protected Areas in India', *Indian Journal of Marine Sciences*, 32(3): 226–33.

Sridhar, A. 2011. 'Precaution in Coastal Regulation: From Principle to Practice', p. 30. Bengaluru: Dakshin Foundation.

Subhash Chandran M. D., T. V. Ramachandra, N. V. Joshi, et al. 2012. 'Conservation of Mangroves of Uttara Kannada District', *Envis Technical Report 50*. Available at http://wgbis.ces.iisc.ernet.in/biodiversity/pubs/ETR/ETR50/ETR50.pdf (accessed 24 February 2017).

Times News Network. 2015. 'Kumta Fishermen Oppose Tadri Port Expansion', *The Times of India,* 21 March 2001. Available at https://timesofindia.indiatimes.com/Kumta-fishermen-oppose-Tadri-port-expansion/articleshow/34959968.cms (accessed 31 July 2015).

7

Sustainability of Endemic Chilgoza Pine Forests in the Western Himalayas

Habitat Threats and Conservation Exigencies[1]

RINKI SARKAR

INTRODUCTION

The motivation behind this study is entirely accidental and a result of my frequent field visits to Kinnaur, a region which spans across the higher reaches of the Indian Western Himalayas in Himachal Pradesh. The steeply undulating terrain in close proximity to perennially snow-capped peaks and the precariously perched mountain villages invariably triggered my lust for wilderness and heights. As a result, I found myself being irresistibly drawn back to this mesmerising landscape to gratify my exploratory cravings.

On one such occasion, while ambling along the alleys of a village named Kalpa, I chanced upon a group of locals huddled around a wooden pole. Adorned in striking customary unisex green velvet caps, they appeared to be peering over a hand-written notice of sorts. I mingled into the crowd quite effortlessly in a bid to decipher the contents and found that the displayed sheet of paper contained a list of names, neatly arranged in a certain sequence. A few friendly gestures and queries revealed that the series related to rotational irrigation water distribution regimes across households in the village. The notice was signed by the *Gaon Vikas* Committee (Village Development Committee), apparently an indigenously constituted governing body that has been in existence for a very long time. The emerging scenario aptly resembled Robert Wade's much acclaimed field study narratives of successfully governed irrigation water cooperatives in the villages of South India.[2]

My research instincts prompted me to enquire if other forms of common property resources were also being collectively managed in this manner. This is when I stumbled upon a local edible pine nut known as *neoza* or chilgoza, which is collectively extracted from the wild by the native population for self-consumption and sale. The resource is essentially a seed embedded in the cones of a specific pine tree that these mountain inhabitants were referring to as *Ree*. Until then, I could barely distinguish one pine from another. My perceptions of the temperate flora were restricted to generic categories such as conifers or pines. Consequently, I was totally oblivious of this distinctive tree species' existence in the wilderness of these elevated tracts. A spell of research-probing revealed that this species of pine is scientifically known as *Pinus gerardiana* to hail the indefatigable spirit of British explorer Captain Alexander Gerard, who defied perilous topographical barriers to penetrate this far-flung and relatively secluded region, way back in 1817. In his travelogue, Gerard vividly described the landscape to be rugged and mountainous to an extraordinary degree.[3] He was also the first to spot the obscure pine, introducing it as a hitherto unknown species within the formal domains of the botanical world. I gathered that this was indeed a rare variety of pine and its distribution was limited to scattered locations in the north-western Hindu Kush Himalayas. This unfolding sequence of revelations was indeed captivating. Garbed in my exploratory research hat, I lost no time to set off on observation trails across the species' native habitat.

Preliminary field impressions propelled my thoughts on to an important ecological concern. If the seeds were being over-extracted for commercial reasons, would it not threaten the natural regeneration and long-term sustainability of this range-restricted species? However, would local governance mechanisms of resource use, which have been in place for a long time, mediate any such unsustainable trends? After all, the local community had a stake in the resource. There I was in an arena that needed me to assess and weigh the ecological versus livelihood outcomes of natural resource exploitation. Except in this case, conservation concerns appeared to be of prime importance as the temperate pine species under scrutiny was rare and endemic to small, fragmented landscapes of these mountain domains (see Image 7.1).

Image 7.1: Extracted Chilgoza Pine Cones

Source: Photo-documentation during preliminary observation trails in 2008.

On my way back to Shimla from Kinnaur, I decided to approach the custodian of these chilgoza forests, namely the State Forest Department, for discussions. I felt it was imperative to express my intuitively tenable apprehensions regarding resource over-extraction and the exigencies of prioritising the conservation of an endemic species. I did not hesitate to propose the urgency of a field-based research study for comprehending and evaluating the gravity of the situation. My projected persistence and genuine concerns did not go unheeded and a study was commissioned by the State Forest Department. What began as an accidental encounter, therefore, turned into a full-fledged field research endeavour for unravelling and assessing the interactive realms of a unique and relatively unexplored forest ecosystem of the Himalayas, intermeshed with livelihood linkages.

The study, initiated in 2009, was purposefully designed to encompass the entire expanse of the chilgoza pine forest habitat of around 2,000 hectares across the mountainous tracts of upper Kinnaur, between 2,000 and 3,000 metres above mean sea level

Figure 7.1: Field Study Area and Location of Case Study Villages

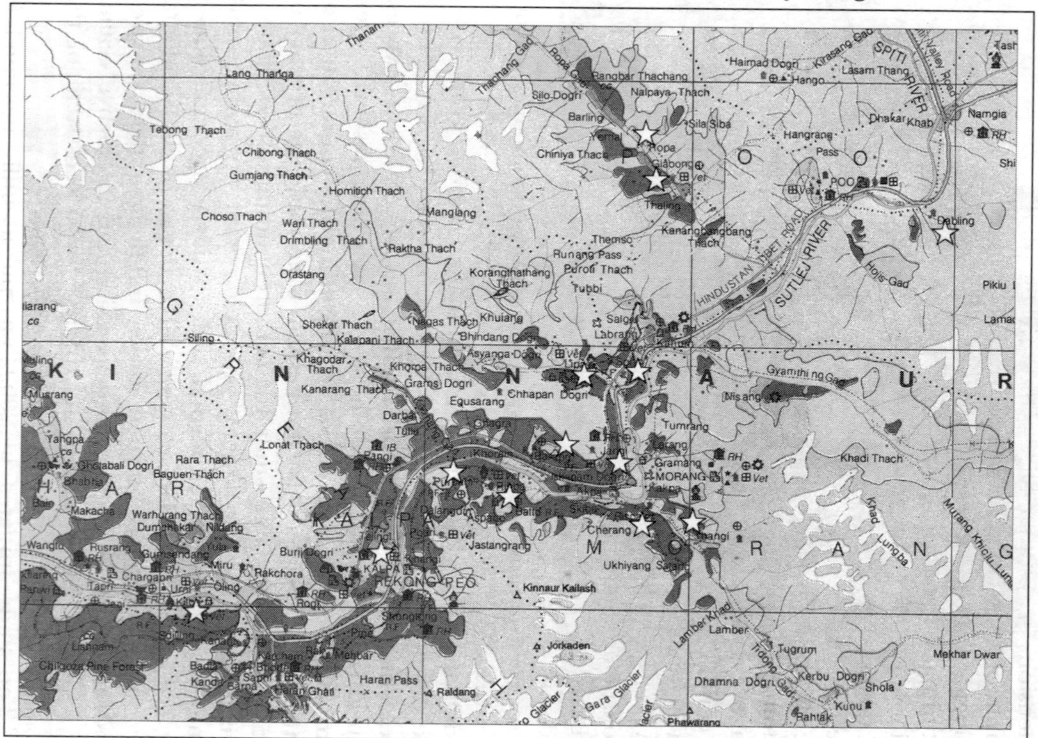

Note: The dark grey patches represent mixed chilgoza forest.

Source: Adapted from the District Planning Map Series, National Atlas and Thematic Mapping Organization (NATMO), Kolkata, India.

(Figure 7.1). A large part of the study tenure was intentionally chosen to coincide with the time period when pine cones are seasonally extracted so that I got ample opportunity to observe and interpret various facets of the institutional as well as physical dimensions of resource access and use regimes. A representative sample of 13 villages was selected for study within the stipulated ecological zone to ensure sufficient variability in aspect, access and altitudinal parameters in order to adequately reflect the natural and topographical dimensions of the region (see Figure 7.1). Care was taken to ensure that the entire expanse of the chilgoza pine forests was extensively covered for the purpose of this multifaceted interdisciplinary field study. Since every village has its own customarily designated chilgoza forest patch that it can access, each village and its surrounding forests formed the basic sampling unit for conducting the socio-ecological study. In all, 257 households[4] and 59 forest stands[5] were assessed through structured quantitative surveys as well as qualitative open-ended ethnographic studies for a holistic understanding of nature–society interactions.

Based on an amalgamation of evidence emanating from these field studies as well as selective archival resources, this paper traces the socio-ecological trajectory of chilgoza pine forests by unfolding the dynamics of a rapidly changing human-inhabited mountain landscape and its impact on resource use regimes, livelihood stakes and the status of the exploited chilgoza pine forest resource base. The consequences of altered resource use regimes for the natural regeneration of this endemic tree species and its long-term sustainability are specifically addressed and evaluated. In light of vital insights emerging from this study, the concluding sections delve into the overall efficacy of the existing conservation environment while recasting thoughts on the usually accepted tropes surrounding the environment–development discourse.

Socio-ecological Dimensions of the Chilgoza Pine

The global distribution of chilgoza pine forests is extremely fragmented and sparse. These forest stands span localised arid pockets of eastern Afghanistan, small tracts of Balochisthan and Waziristan in Pakistan, as well as scattered locations in the upper reaches of the Chenab, Ravi and Sutlej river valleys of the north-west Indian Himalayas. The largest expanse of chilgoza forests in

India is to be found in Kinnaur district, where the field studies were conducted. In all these areas, the species is a local endemic,[6] occupying mostly steep xeric mountain slopes with elevations between 1,800 and 3,500 metres.

Taxonomically, the chilgoza pine belongs to the *Pinus* genus and has a white or soft pine lineage,[7] narrowly classified under the sub-section *Gerardianae*. This sub-section, which includes two other affiliated species from East Asia,[8] is a small distinct group of relict species, identified by their characteristically smooth bark that exfoliates into large, thin scales, giving them a mottled appearance.[9]

The apparently patchy occurrence of chilgoza pine forests was puzzling for me, and I immersed myself into a flurry of research investigations that yielded striking revelations. It appears that the *Gerardianae* pine lineage has meagre fossil records.[10] Therefore, there is scanty literature on the biogeographic[11] history of *Pinus gerardiana*. However, more recent phylogenetic studies conducted for reconstructing the ancestral lineage of this white pine group indicate that unusual geological and climatic events in the Cenozoic geological era may have influenced the present distribution and evolutionary patterns of these temperate species.[12] For instance, oscillating climatic conditions during the Pleistocene glaciation and active mountain-building during the late Neogene period seem to have triggered isolating environments, causing species divergence.

More specifically, these studies suggest that the common Mediterranean ancestral population of the subsection *Gerardianae* was dispersed along the erstwhile Tethys coast towards the eastern to the south-eastern edge of the Quinghai Tibetan Plateau. Uplift of this plateau, during the late Tertiary period, created major physical and ecological barriers that segregated the ancestral population of *Gerardianae* into isolated enclaves by high mountains, preventing genetic exchange. Under varying environmental conditions, these eventually became differentiated into distinct species.[13] The peculiar isolated occurrence and non-contiguous patchy distribution of chilgoza pine forests, observed across the Hindu Kush Himalayas, can be attributed to these factors, a phenomenon also known as 'geographic speciation'.[14]

These segmented Chilgoza pine forests are native to stressful high montane subalpine environments that are water-constrained or very windy, with long, dry, cold periods and a short growing season. Precipitation is mostly in the form of snow. This slow-growing[15] and long-lived pine species[16] is a hardy one well-adapted to these constrained surroundings. For instance, it has the ability to bear severe cold in winter as well as excessive drought. Further, the trees are wind-firm and make little demand on the fertility of the soil.[17] All along the field research study area, clusters of chilgoza forest stands can be found on a variety of mountain topographies ranging from gentle barren rolling terrain to rocky ridges as well as steep rocky slopes. Such ecological attributes endow these forests with the ability to perform vital hydrological and soil-stabilising functions for sustaining the mountain ecosystem that is inherently susceptible to flooding and erosion caused by wind or the down-slope movement of water or snow.

The entire expanse of chilgoza forests across Kinnaur is restricted to the semi-arid ecological belt between the small local hub Wangtu and a little beyond the settlement of Pooh on either side of the Sutlej River. Towards Wangtu, where the influence of the monsoon does not wear away entirely, these forests are mingled with the stray occurrence of holm or holly oak (*Quercus ilex*), known to be native to the Mediterranean region.[18] *Cedrus deodara* trees are also encountered in the upper periphery. Beyond this belt, towards the northeast, the oaks disappear as the landscape becomes more and more arid, typically resembling 'Himalayan dry temperate forests'[19] (Figure 7.1). In this ecological zone, chilgoza tracts are mainly associated with scattered *Cedrus deodara*, some *Fraxinus xanthoxyloides* and *Juniperus spp.* as the elevation increases. The understore is mainly dominated by innumerable *Artemisia spp.*, while *Astralagus spp.* and *Daphne spp.* are found in some quarters. The last traces of scattered stunted chilgoza trees are visible on steep slopes above the confluence of the Spiti and Sutlej rivers. Beyond this juncture, the landscape assumes a bare and barren appearance almost devoid of any vegetation.

The chilgoza pine tree has distinct characteristics in comparison to other evergreen needle-leaf conifer species of the temperate Western Himalayas. It can be easily distinguished from these species because of its exfoliating bark, which displays a greenish-grey scaly appearance (see Image 7.2). The tree usually attains a height between 15–20 metres and a girth of 2–3 metres, but the overall morphological structure of the tree varies across the mountain landscape. Large crowns and twisted ascending branches are visible in open areas, while moderately dense slender poles are common to closed canopy stretches. In some areas, multi-stem growth is also observable. At very high altitudes, as the landscape turns arid, assuming cold desert-like conditions, the trees tend to be more stunted and sparse in occurrence.

Besides these broad morphological dimensions, the chilgoza tree has other unique traits. For instance, the leaves are long and needle-like in bundles of three. Pine cones which tend to dangle from the edge of a branch are also very peculiar, resembling a small green pineapple or an artichoke.[20] These relatively large-sized cones are nearly indehiscent and take about 17 months to ripen fully after pollination. The long slender seeds embedded in these cones are wingless and are fairly large in size, loaded with nutrients (see Image 7.2).[21] The large seed size is considered to be an adaptation to the stressful xeric environmental conditions.[22] The existence of a short growing season makes the large size advantageous as these seeds can then provide an adequate and timely supply of nutrients, necessary for rapid early growth and establishment.

Image 7.2: Chilgoza Pine Leaves and a Cone

Source: Photo-documented in course of field surveys, 2009.

All species of pine trees nurture seeds that are edible. However, the relatively over-sized dimensions of chilgoza pine seeds make them amenable to easy extraction and worthy enough for human consumption. In Kinnaur, the extraction of pine seeds or pine nuts from the wild by the native population predates recorded history.[23] Field-based narratives pertaining to the past revealed that the harsh terrain limited cultivation possibilities to a meagre range of crops with low productivity levels. Therefore, pine nuts supplemented dietary needs, particularly during winter months. Because of these livelihood-linked necessities associated with chilgoza forests, the local population has always had customary rights to harvest pine nuts from state-owned forests in the village vicinity.

Chilgoza pine cones are usually ready for collection during late September. Menfolk climb trees to great heights, detecting and lopping branches laden with cones. These cones, strewn across the forest floor, are then accumulated and transported to cattle pens. Here they are stored under a thick layer of stocked grass fodder for about a fortnight to infuse warm conditions, which makes it easier to break them open. Each seed is then meticulously removed by hand. Therefore, the entire activity from the extraction of cones to the removal of seeds for consumption is extremely tedious, involving considerable human toil, skill and dexterity. In many locations, the precarious nature of the terrain makes this venture risk-prone as well.

The native population primarily values chilgoza forests for the edible pine nuts these trees yield. However, these forests have been an integral part of other kinds of survival exigencies and ingenious ways have been devised to tap the utility of these pine trees. For instance, as the wood is very resinous,[24] wood-chips extracted from these trees were the only effective source of torchlight when the region was devoid of electricity coverage even until the late 1960s. These pine torches, known as *jogti* in the local dialect, were of great assistance for monitoring the rotational supply of irrigation water across households during the night. In the course of group discussions, some exuberant elderly respondents hailed how the *jogti* was the only means for overseeing the continuous operation of locally designed water-mills installed for grinding local grains to flour. After all, adequate food security was crucial to overcome hardships during the long and formidable winter season. Although the chilgoza pine is a mediocre source of timber, in sparsely vegetated pockets that are not rich in preferred timber species, there is evidence of this pine being resorted to for construction purposes.

Perhaps because of its ubiquitous presence in the vicinity of villages and for the multiplicity of livelihood functions these forests perform, it was awe-inspiring to gather that the chilgoza tree is actually revered by the local community and considered to be auspicious in ritualistic and cultural spheres. In a region where floral wealth is meagre, particularly during winter months, pine seeds are set aside by every household for making garlands as offerings to local deities or to welcome guests as a gesture of hospitality.

During festivities such as weddings, it is obligatory for every household to gift strings of pine seeds to the host family. I gathered that this unusual ritual was a symbolic collective venture for acknowledging the sanctity of the solemn occasion. While performing last rites in times of death, chilgoza trees are the only ritually accepted source of cremation wood in this region to fuel the burning pyre (Image 7.3).

Image 7.3: Customary Uses of Pine Nuts in Ritualistic Spheres

Roasting edible pine nuts

Pine nut garland offerings at a wedding

Strings of pine nuts ready for ritualistic offerings

Pine nut garland offerings: A sign of hospitality

Pine nut garland offerings to a local deity

Source: Field studies 2009.

The socio-ecological dimensions highlighted in this section amply substantiate the significance of ecosystem services provided by chilgoza pine forests to sustain an environmentally stressed habitat. My field research endeavour also exemplified the deep-rooted interface between a native society and a well-adapted pine species that has developed to cope with livelihood hardships in a difficult geographical domain. Given these reinforcing nature–society interactions that I had unravelled while sharing the hearth and lively companionship of my expanding native affiliations, I wondered whether my fears about anthropogenic threats to this rare species were justified. Perhaps I needed to engage myself in a deeper scrutiny of resource use and its impact on the resource in the light of changing contextual parameters.

Regional Dynamics: Narratives of Transformation and Change

Ethnography of Pre-Modern Times: Prior to 1960

Due to topographical barriers, Kinnaur remained relatively secluded and sparsely populated for a long spell of time. Although there was some seasonal transhumance movement of sheep and goat flocks into pasture lands of neighbouring Tibet and occasional visits to the royal precincts of the Bushahr dynasty in the riverside town of Rampur, the region remained completely cutoff, particularly during winter that stretched for a large fraction of the year. During this bitterly cold period, the local inhabitants were mostly confined to their village homes due to heavy and persistent snowfall. Consequently, inter-regional mobility was meagre and restricted. However, movements across neighbouring villages were aided by narrow foot trails that crisscrossed the steep mountain slopes.

Early nineteenth-century travel narratives of British explorers provide vivid descriptions of the treacherous, rocky and steep landscape, invariably conjuring up memories of unknown and inaccessible valleys.[25] They reiterate that overcoming these frequently intercepting hazards along undulating mountain trails was a sheer test of one's endurance and agility. Unstable and dangerously narrow foot trails along a steep precipice were the only paths for mobility, '... giving the traveller uncertain idea of his safety'.[26] It was much later, in 1850, that the first road alignment of sorts through the region was commissioned by the British regime in the hope of fostering trade with the adjacent territories of Tibet. Designated as the Hindustan–Tibet road, the elevated forest path was proposed for administrative reasons as well, to improve accessibility conditions necessary for governing these far-flung areas. Even until the early 1960s, the carved mountain path continued to be a narrow unmetalled mule-track that was completely blocked for access during winter due to heavy accumulation of snow. Subsequently, the British found it uneconomical to upgrade the bridle path to a motor road due to the small amount of trade it fostered and frequent disruptions in winter.[27] As a consequence, the region continued to be largely inaccessible.

Oral history accounts of the past, gathered in the course of fieldwork, enunciate great hardships borne by the native population to cater to their livelihood necessities. The xeric landscape and extreme environment curtailed the growing season to a very short period, permitting only a few varieties of millets, mostly buckwheat (*Fagopyrum sp.*) and barley, to be grown. However, the subalpine landscape was ideal for pastoralism and the locals reared large flocks of sheep and goats, which were known to be the principal source of their wealth and riches.[28] To cater to their day-to-day existence, these mountain dwellers devised ingenious ways for harnessing natural resources besides adapting to labour resource constraints due to the meagre population base in an isolated domain. As a result, a high degree of cooperative spirit and reciprocating social arrangements, known as *sckol* in local parlance, evolved spontaneously to overcome these difficulties and engagements that required great human labour and toil.[29] All across the region, these social systems were reinforced through effective authority structures under the surveillance of powerful local deities believed to cause harm to those obstructing social harmony.

The harvesting of chilgoza pine nuts also displayed high levels of collective action guided by well-formulated rules of local governance, framed to ensure fair investment of labour as well as equitable access to the best quality produce from these common forests. For instance, when the pine cones were ready for collection, the forests were barred from entry until a pre-determined date was declared by the local *devta* (deity) Committee to prevent unfair surreptitious extractions. When the forests were finally opened, it was mandatory for each household to participate in cone collection activities. On the stipulated day, one female and one male member from each household were permitted to engage in resource extraction. The male members took the lead in climbing trees and '... by means of a long pole with a hook at the end, they wrenched off the cones, breaking a piece of the branch.'[30] Their female counterparts were entrusted with the task of collecting cones that tumbled onto the forest floor. All the cones harvested by the village community in this manner were then pooled together in one spot and distributed equally across households. Informal monitoring and fear of social opprobrium ensured active participation and rule compliance keeping violations at bay. Thereafter, the forests became open access to harness cones that may have been left behind.

Field interviews with the aged members of the surveyed villages confirmed that extraction levels were limited in the past and it was quite common to find mature cones dangling from the trees long after the collective cone extraction spell was over. They reiterated that demographic pressure on the resource was meagre at that time and pine nut needs were limited to self-consumption requirements, mainly while a small quantum of nuts was stocked for making ritual garlands. Further, overall cone collection beyond these subsistence levels was virtually negligible due to time constraints as this period coincided with other crucial labour-intensive engagements such as harvesting followed by stocking of cultivated crops as well as firewood, fodder and leaf litter from forests to prepare for the long winter season when locals were forced to remain confined to their homes due to extreme weather conditions.

But more significantly, there was barely any pine nut collection for sale during this era. Gerard's local encounter with the natives in the early part of the nineteenth century provides crucial evidence on the virtual lack of pine nut marketability in Kinnaur and the consequent low value of the produce. In his travel diary, he narrates that he was charged only one rupee for 30–36 pounds of pine nuts in this remote belt, while one rupee fetched only one pound of pine nuts in the town of Moradabad at that time, for the resource extracted from chilgoza forests in Peshawar.[31] He deduced that only a small quantum of the resource found its way to the plains.

Gerard's observations appeared to reinforce local accounts of the past gathered in the course of field work. The Kinnaur pastoralists usually traded wool and raisins processed from their local vineyards at the entrepôt town of Rampur-Bushahr in exchange for grains from the plains. A small quantum of pine nut and cumin, extracted from the wild, occasionally featured in these trade dealings. Sheep and goats were the only available forms of baggage animals that were capable of negotiating the narrow mountain trails. Vivid local narratives recounted that the arduous journey to the plains and such feeble modes of transportation prevented the harvesting of pine nuts on a commercial scale.

Pre-modern society, in essence, portrayed a scenario characterised by self-sustaining livelihood systems based on adaptation and collective strategies for coping with an inhospitable terrain and secluded existence. As a consequence, the market potential of the pine nut yield remained untapped. Under this set of circumstances, it may well be presumed that the natural resource base must have been much in excess of the extraction levels, with meagre threats to natural regeneration. However, the colonial British forest evaluators who ventured into these forests during the early part of the nineteenth century did lament about the abrasive nature of the extraction. They observed that much damage was done to the trees during cone collection activities.[32]

The Changing Landscape

The post-1960s era has ushered in a period of dramatic transformation for the entire expanse of the Kinnaur district. The speedy completion of National Highway 22, following the 1962 Indo-China war, has been an important driving force triggering these new developments (Sarkar 2010). The highway meanders along the Sutlej River and cuts through core tracts of the district. Conceptualised as an alternative route, the old colonial Hindustan–Tibet road alignment was mostly abandoned and the National Highway was designed to follow a different course at lower elevations to ensure all-weather road connectivity. Villages on either side of the mountain slopes are now well-connected to the Highway by a network of rural link roads that have come up in more recent decades due to the Government's concerted efforts to promote regional development through better road connectivity. Although the National Highway was conceived primarily to cope with strategic national security compulsions, it inadvertently opened up an inhospitable and isolated region, with far-reaching socio-economic and environmental consequences.

For instance, local perceptions captured in the course of field interactions have consistently emphasised the role of this

Highway in boosting horticulture development by facilitating the transportation of apples to distant markets where the produce fetches high returns.[33] It appeared that even before the 1960s era, some local visionaries with foresight were eager to pursue apple cultivation because in the course of their travels outside the region, they were exposed to lucrative outcomes of the Kotgarh horticulture revolution initiated by an American philanthropist named Stokes, in neighbouring belts in Shimla district.[34] However, Kinnaur's remote bearings dissuaded them from embarking on a new venture of this nature, which also entailed a long and risky gestation period. The coming up of National Highway 22 unshackled these barriers for realising potential economic gains by facilitating a steady crop shift towards apple production.

A host of other favourable factors have hastened the pace of transition towards horticulture. The temperate climate that is extremely suitable for growing good quality apples, the possibility of exploiting age-old techniques of tapping water from seasonal glacial snow-melts for keeping orchards well-irrigated, adequate landholding for diversifying agricultural production towards an alien crop, and the well-nurtured entrepreneurial abilities of the Kinnaur people, due to their ancient trans-montane trade with Tibet, have all been crucial enabling factors aiding this tilt towards the preference for commercial apple cultivation.[35] As a consequence, there has been a radical transformation in cropping patterns. In 1970, nearly 85 per cent of the total cultivated area was under traditional mountain millets in the region.[36] However, household surveys conducted in the representative sample of surveyed villages revealed that in the present context, nearly 90 per cent of a household's cultivated landholding is under apple orchards. This truly remarkable and rapid pace of transition was found to be consistent in all villages and across all factions of society. Due to these emerging trends, at the aggregate district level, the cultivated area under apple crops has increased phenomenally by 800 times in five decades from 13 hectares in 1961[37] to 10,828 hectares in 2015.[38]

Better road connectivity has also been a crucial factor for promoting access to higher education that has played an important instrumental role in enabling recruitments to government jobs. Income from these non-farm avenues of employment supplements returns from horticulture as the local population has judiciously exploited the government job reservation policy granted to them by virtue of their special tribal status.[39]

These sustained occupation shifts towards horticulture and non-farm sources of income have had a tremendous impact on the economic condition of the local population. Although there seems to have been no evidence of destitution even in the past, as Gerard in his travel impressions of Kinnaur district in 1817 succinctly remarks:

> ... The *Koonawarees* are all traders and their chief riches consist in large flocks of sheep and goats that furnish them with wool which together with raisins they exchange for grains .... Notwithstanding the want of grain there are no marks of poverty in the country ... for even if grain be scarce, as it often is, yet their large flocks furnish an inexhaustible store .... The people are as well clad as they can be and seem to enjoy a much greater degree of comfort than any of the other mountaineers I have seen.[40]

Nonetheless, the present emerging trends are indeed spectacular, reinforcing relatively high levels of economic well-being the region has exhibited over time (Figure 7.2). A comparative quantitative assessment of rural income structures based on

Figure 7.2: Levels of Economic Prosperity in Kinnaur

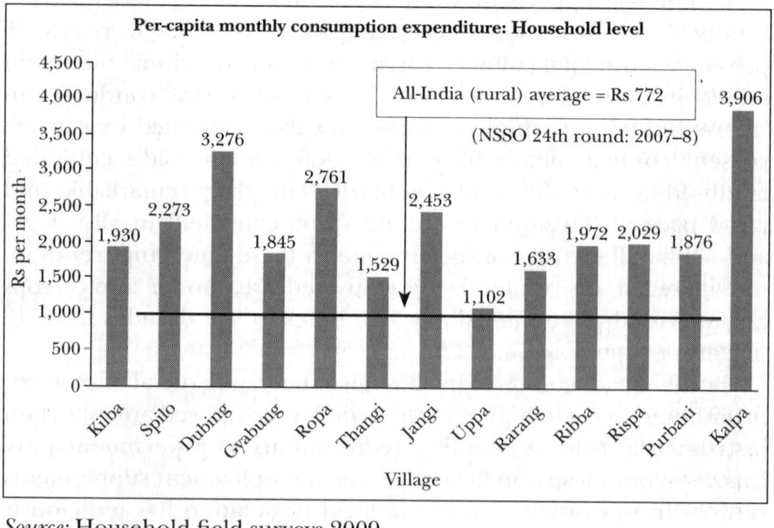

Source: Household field surveys 2009.

household surveys in each of the 13 case study villages indicates that the levels of economic prosperity of the local population, assessed in terms of the per capita monthly consumption expenditure, is way above the national average for all of rural India (Figure 7.2). These trends have been fairly inclusive and poverty is a non-issue in the region. Unfortunately, the effects of these overall developments have not been so benign for chilgoza pine forests in the region, as my research findings suggest.

Impact of the Changing Scenario on Chilgoza Pine Forests

Due to the changing façade of this Himalayan landscape triggered by better road connectivity and market integration, the sale of chilgoza pine nuts has become a very profitable proposition for the local inhabitants. A review of relevant literature and supportive field narratives confirmed booming demand conditions for these pine nuts in national as well as international markets.[41] As the resource originates from this restricted geographic domain,[42] the produce fetches a high price at the local level, escalating many more times until it reaches urban retail centres in the plains. The mounting lure of discernible lucrative gains, hitherto unknown in the past, has progressively induced *reckless and near-total harvesting of pine cones* for the market. Trends emanating from the household surveys indicate that 85–90 per cent of the pine nut yield is earmarked for sale, and the remainder set aside for self-use. This is in sharp contrast to earlier times when most of the crop was kept for self-consumption. To cope with the unfolding circumstances, heavily biased towards horticulture-oriented rural development coupled with rising cone collection exigencies, traditional pine nut resource use regimes and extraction mechanisms have also altered radically.

First, the opportunity cost of engaging in old practices of collective cone collection has risen significantly due to shifting occupational priorities. This is because the period when cones are ready for harvest tends to coincide with the peak apple harvesting season. Therefore, most inhabitants are hard-pressed for time to participate in mandatory social arrangements surrounding cone extraction activities. Further, a large fraction of the able-bodied population is also away pursuing time-bound government jobs or educational pursuits within and outside the region. Second, as economic prosperity and comfort levels have risen over the years, village households are now less inclined to engage themselves directly in the tedious activities related to pine nut collection.

The burgeoning pool of migrant casual labourers, who throng the region to secure employment in orchards and a host of other load-bearing activities, have become an important resource for overcoming such non-participation woes. Hailing from similar mountain tracts of Nepal, they are well-versed to shoulder the hardships of working in a harsh undulating terrain. The burden of pine nut extraction is mostly shouldered by them in the present scenario. Amicably addressed as *bahadurs* in the region, these migrant employees are paid a higher wage as an incentive to ensure their undeterred involvement in tasks that are much more dangerous now, as forests even on steep slopes need to be tapped for reaping maximum pine nut yield. Besides, they need to be adequately compensated as pine cone collection has to be done with greater speed and dexterity for timely dispatch of the produce to distant markets in the plains.

New resource extraction arrangements have also evolved through mutual consent so that it is no longer essential for every household to be present at the designated cone collection venue. For instance, in about half the villages that were surveyed, group-access regimes have substituted older norms of resource extraction. The remaining villages have started auctioning their chilgoza pine forests in order to abstain from cone collection altogether.

In the set of villages that rely on group-access methods of extraction, households are formed into groups and their forests are divided into equal-sized compartments. Each group is assigned one compartment in different locations on a rotational basis every year because the quality and quantity of cone production may vary across the zones. In some villages, a lottery system is resorted to for allocating compartments among the groups. In most cases, migrant labourers are usually engaged for extracting the cones, distributing the yield equally across group members, and transporting the produce back to the village. In case a household is unable to join the collection activities, their hired labourer can act as a substitute to represent them with the prior approval of other group members. Alternatively, they can delegate the charge to perform their role in the cooperative venture to a willing relative or a neighbour.

In the course of field observations, it was invariably found that household members who do manage to take time out to participate usually supervise the extraction operations, besides making sure that the labourers do not erroneously stray into another group's demarcated cone collection area. Much to my dismay, there was virtually no evidence to substantiate that their regulatory actions were also targetted at ensuring sustainable extraction so that less damage was caused to the trees by the labourers who seemed to be indulging in ruthless lopping of branches to secure the cones.

In villages where the scale of apple cultivation is on the higher side, the locals have judiciously initiated mechanisms to auction out their chilgoza pine forests collectively to local *thekedars* (contractors or middle-men). It is a win–win scenario for both parties entering into this new arrangement. Village households are relieved to delegate the entire task in this manner as they prefer to focus on the management of their apple produce in an unhindered way. Monetary receipts from the auction are either equally distributed across the households or directly siphoned away to replenish the local deity's temple treasury, by common consent. The auction proceeds are meagre compared to what a household could have earned by actively participating in cone collection and sale. But they are willing to forgo these returns as their income from apple sales more than compensates for this minor loss. So they channelise their time and effort towards nurturing financially rewarding horticultural pursuits. It pays for the auction winner to bid for exclusive rights to the valued pine nut resource as the profit margins from sale in urban markets are quite substantial. The auction winner, however, has to bear the risks of swift transportation of the resource to the wholesale centre for best returns.

All in all, these field insights across chilgoza pine forests reaffirmed that locals have started distancing themselves from the natural resource and possibly, from the resource condition as well. These trends were indeed disconcerting and distinct as compared to narratives of resource access and use in the past; and I seemed to be edging closer to confirming that my preliminary apprehensions regarding imminent threats to a rare and endemic species of pine were not ill-founded, after all.

My research endeavours subsequently established that concomitant to the changing socio-economic and institutional context, the *pace of pine nut extraction has also increased.* I deduced that this was due to other impending time constraints, which were irrelevant in the past. For instance, the produce has to be extracted swiftly so that it reaches urban markets before the onset of the festive Diwali season when demand tends to peak. Apparently, pine seeds/nuts tend to lose moisture very rapidly and any delays can reduce the quantum of resource to be sold in weightage, resulting in potential income losses. Therefore, there is undue pressure to harvest quickly. The faster the speed of extraction, the greater is the chance of procuring a substantive margin over costs by the time the produce reaches the market. Homeward-bound migrant labourers are also in a frenzy to complete extraction activities because many of them wish to reach their native places before the upcoming festive season with their kitty of earnings. The impact of these compulsions for speedy extraction have been quite alarming for chilgoza pine forests as old techniques of pine cone collection that were less harmful have been substituted by far more damaging ones, resulting in *excessive and unsustainable lopping of branches.*

Archival referencing and field interviews revealed that in the past, a long pole with a sharp iron hook at the end was used to wrench off cones from the trees.[43] I was lucky to find samples of the tool which some households had retained in their stores. They were kind to demonstrate its usage, which confirmed that the traditional methods of extraction caused relatively less damage to the tree (Figure 7.3). However, local inhabitants asserted that the older extraction tool has been largely abandoned as it was too unwieldy and slow for the present circumstances; more ground has to be covered at a much faster pace of cone collection to facilitate rapid and near-total removal of cones for optimum economic gains.

With mounting commercialisation pressures, the locals have adopted a handier tool known as *drati,* which has definitely enhanced the speed of cone collection operations, but with dire consequences for the forests. British forestry experts from the colonial era did lament about injury caused to the trees due to cone collection, even in the past.[44] However, threats to the resource must have been less severe due to meagre extraction levels. The magnitude of extraction has certainly increased by inducing these destructive mechanisms in the present context.[45] All across the field research areas, I observed that the intensity of lopping and the pace at which trees were being scaled by cone collectors left the trees badly mutilated (see Figures 7.3 and 7.4).

Figure 7.3: Comparative Depiction of Extraction Mechanisms in the Past and Present

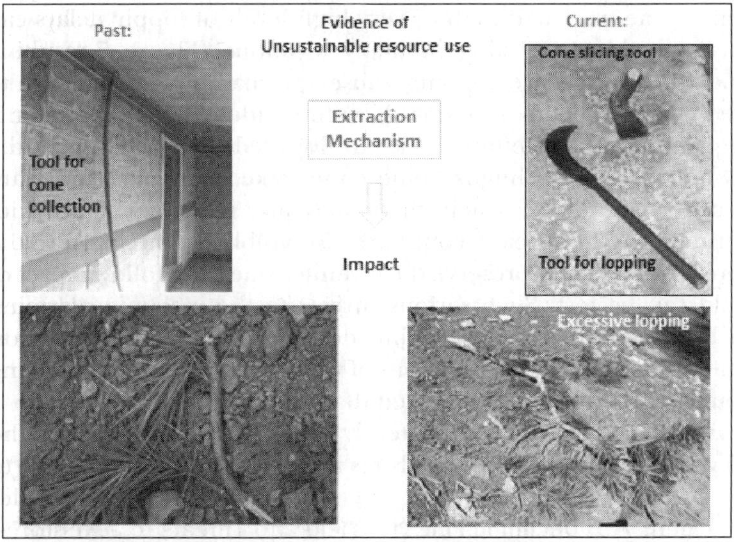

Source: Field studies 2009.

Figure 7.4: Visual Evidence of Tree Damage Caused by Lopping for Cone Extraction

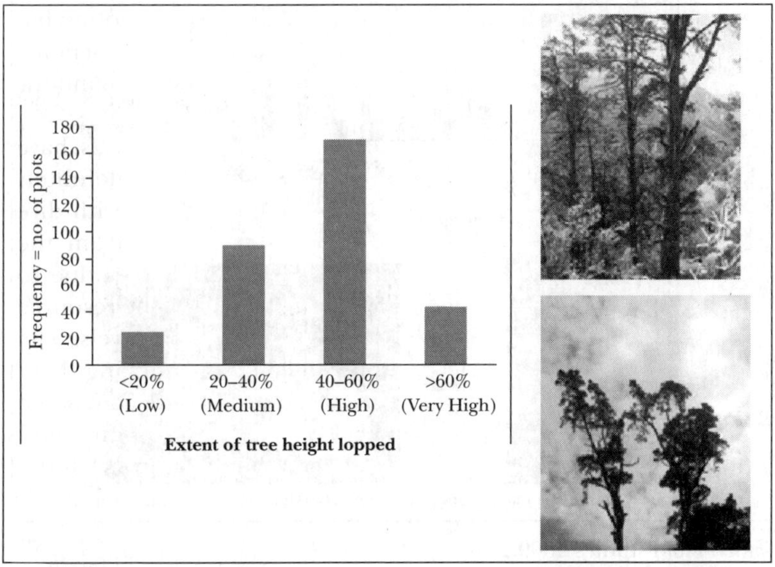

Source: Field studies 2009.

Corroborative quantitative evidence emerging from data collected in chilgoza pine forests indicated that nearly 65 per cent of the trees had high or very high levels of lopping, showing unsustainable harvesting in many locations (Figure 7.4).[46] For one thing, excessive lopping causes permanent damage to the tree's overall metabolism and growth, besides thwarting potential regeneration possibilities due to lowered cone production.[47] As noted earlier, Chilgoza pine cones take nearly 17 months to mature. Therefore, when the cones are ready for harvest in autumn, the next year's cones are also visible on the branches. If care is not taken to preserve the juvenile cones, the following year's yield can be lost. Such serious threats to the resource are likely to be important reasons behind declining pine nut yield levels. Inter-temporal local perceptions of yield levels, documented in the course of field studies, indicated that, but for the periodic 'mast'[48] years, yield levels have consistently declined by half over the last 25 years. This is true for all villages in the study area (Figure 7.5).

Figure 7.5: Declining Pine Nut Yield and Threats to Juvenile Cones of the Next Season

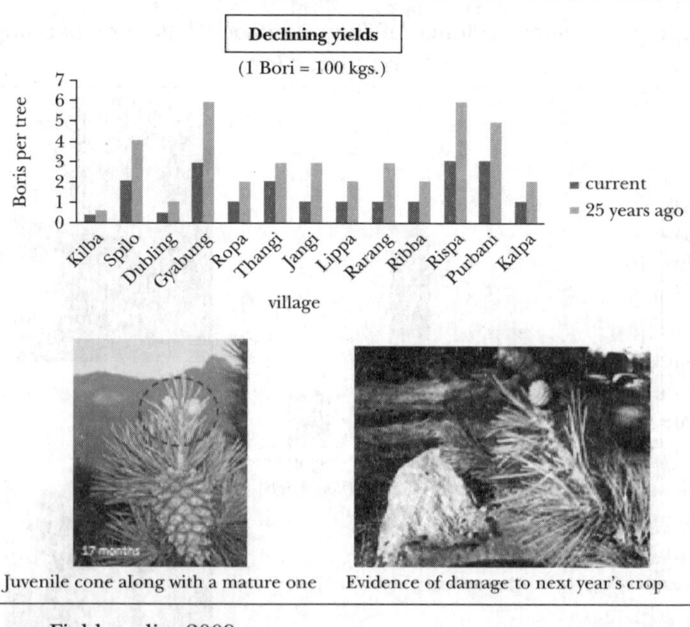

Juvenile cone along with a mature one

Evidence of damage to next year's crop

Source: Field studies 2009.

At this juncture of my research investigation, I thought it would be pertinent to gather the village community's perceptions on the debilitating resource condition. After all, their interface with this natural resource has been deep and prolonged. A majority of my respondents seemed to exhibit a certain sense of myopia to queries about resource status. Swayed by short-term gains, they appeared to be totally oblivious to the adverse ecological consequences of their own actions. Most were boisterous and eager to attribute vagaries of climate change, characterised by subdued winter and warmer temperatures, as the primary cause of declining pine nut yields. They were also unanimous in expressing their concerns over the loss of healthy standing chilgoza forests due to the coming up of major hydropower projects in the region and haphazard road construction. Only a small fraction of my insightful interviewees confided that the new market-oriented regime of excessive lopping for pine cones to cater to extraction for the market, driven by sheer greed, has resulted in declining yield levels. They stressed that these undesirable trends have accentuated with the advent of *thekedars* and the engagement of migrant labour. It is obvious that for these agents associated with pine nut extraction and trade, only the present year's yield matters as their returns are proportionate to the magnitude of their collections. Besides, they are unsure whether such lucrative opportunities will come their way in the following year. So they are least concerned about declining future yield levels, or overall threats to the natural resource and the mountain ecosystem it sustains.

Over the years, these untoward social circumstances have led to a vicious cycle of declining yields, spiralling prices and mounting pressures for near-total collection of cones, jeopardising the condition of these endemic forests. These trends are evident from Figure 7.6, which depicts the impact of market access, rising extraction levels and shrinking yield levels on prices of pine nuts over time.

As my research progressed, I seemed to be just one step away from empirically verifying the impact of excessive resource exploitation on natural regeneration, which is an important indicator for assessing the long-term sustainability of a forest. The commonly used forestry science approach for assessing the status of natural regeneration is to determine the size class distribution of seedlings, saplings and trees in the forests. Typically, an inverse

Figure 7.6: Spiralling Price of Chilgoza Pine Nuts in Kinnaur (1950–2016)

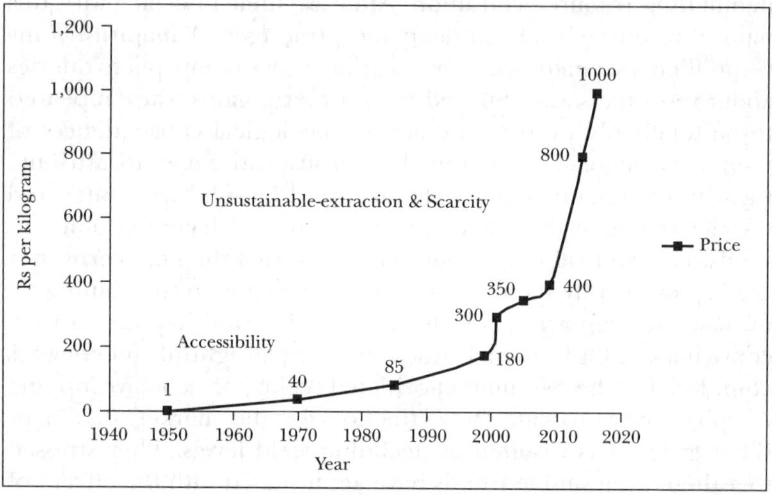

Source: Field studies 2009.

'j-shaped' curve that shows a very high proportion of seedlings and saplings in relation to trees is considered to represent a healthy regenerating population due to the fact that there is some mortality at every size class.[49] The regeneration potential of chilgoza pine forests, based on the requisite data collected through forest surveys reflects an appalling scenario, as depicted in Figure 7.7.

The actual population structure of chilgoza pine forests appears to deviate radically from the ideal scenario that would have ensured the long-term sustainability of this endemic species. It is quite evident that a dismal seedling count of 165 per hectare is insufficient for sustaining the current forest stock. What is worse, there is virtually no sapling establishment in these forests, as evident from the graph in Figure 7.7. Based on visual impressions of near-total cone collection, other researchers have also expressed consequential threats to natural regeneration, much like I had envisaged in the course of my preliminary reconnaissance surveys.[50] But the poor regeneration trends outlined in Figure 7.7 provide quantitative empirical evidence on the gravity of the situation. It is quite obvious from these results that if this scenario continues, the

Figure 7.7: The Status of Natural Regeneration in Chilgoza Pine Forests of Kinnaur

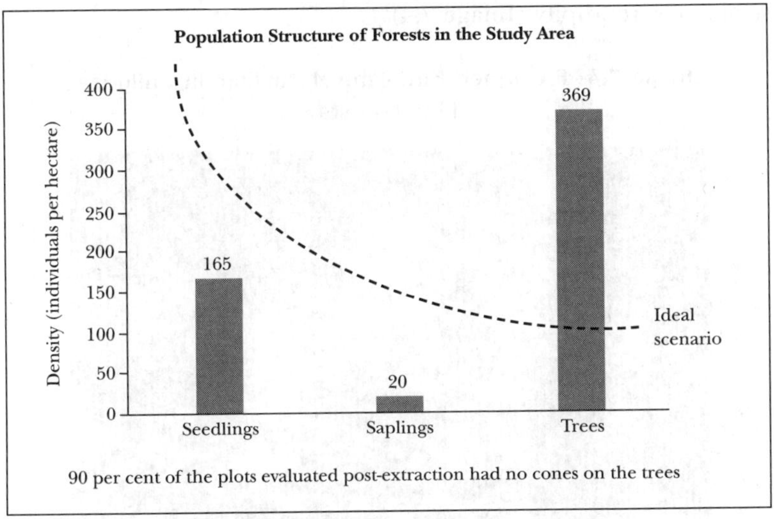

Source: Field studies 2009.

chilgoza trees would eventually age away without a young cohort to replace them.

In addition to the reckless nature of pine cone extraction and the loss of healthy forests to development projects or orchards, these outcomes may also be ascribed to other kinds of biotic pressures such as grazing. Seedlings and saplings are known to be susceptible to damage due to browsing and trampling by livestock. An ordinal scale used to measure the effect of natural and biotic pressures in these forests did point towards moderately high levels of grazing in these zones. A multiplicity of such anthropogenic pressures have perhaps compacted the soil, a state which is detrimental for the healthy growth of recruits. Laboratory tests conducted on soil samples collected from the forests revealed high bulk density values beyond the expected threshold, signifying high compactness or low porosity levels. Such soil conditions are unfavourable for the growth of chilgoza pine seedlings, which requires free porous soil without any excess of moisture.[51] As I delved into the complexities of these untoward circumstances thwarting natural regeneration, my accidental discovery of a crucial biological interaction seemed to provide some solace and hope for the future sustainability of these endemic pine species. The narratives that follow, therefore, depict the triumphs of observational field studies that resulted in a stroke of serendipity (Image 7.4).

Image 7.4: Presumed Bird–Pine Mutualism in Chilgoza Pine Forests

Source: Field work 2009. This greater spotted nutcracker was observed by the author in a chilgoza pine stand in 2009.

In the course of my forest transects, I noticed a lot of healthy regeneration below rocks and wondered how the seeds came to be dispersed in these odd locations. I also spotted clustered seedling growth in some areas, which was unusual. While trailing behind my field team, on a misty morning, I spotted a bird precariously perched atop a mature cone dangling from the branch of a tall chilgoza pine tree. Later, I identified the bird and it turned out to be the greater spotted nutcracker (*Nucifraga multipunctata*) of the corvid family (Image 7.4). My research endeavours revealed that akin to the chilgoza pine, this avian species was also endemic to the Western Himalayas. I learnt that nutcracker species are closely associated with montane coniferous forests across parts of North America, Europe and Asia. While trying to make some sense by piecing my disparate observations together, I was finally drawn to Hutchins and Lanner's research work on the role of avian seed dispersal for pines that have wingless seeds.[52]

It appears that there are 20 species of pines belonging to the genus *Pinus*, which have wingless seeds. Nineteen of these, including the chilgoza pine, fall in the subgenus *Strobus*, which are known to be dispersed by corvids, especially nutcrackers.[53] There is established evidence on this nature of avian seed dispersal in the case of eight of these *Strobus* pines. It is presumed to be present in the rest of the 11 as well (Tomback and Linhart 1990). This bird–pine relationship shows a co-evolved mutualism.[54] While the pine is extremely dependent on its avian seed disperser for its regeneration, the nutcracker is at least heavily dependent on the pine for its primary year-round food source. Studies pertaining to the Western temperate belts have established that these nutcrackers harvest tens of thousands of pine seeds and bury them in small caches for later retrieval during winter, spring and parts of the following summer. Seeds that are not retrieved lie dormant and germinate in favourable years.[55] I deduced that the cropping up of single or multi-stem seedlings in most unusual locations that I had repeatedly spotted, in the course of my forest observation trails, could well be untapped nutcracker caches that had germinated into seedlings.

More specifically, the studies conducted in the temperate spheres of North America and Europe have indicated that nutcrackers are primary seed dispersers for pines that have wingless seeds and therefore *play a crucial role in promoting natural regeneration*.[56] Although some other vertebrates may occasionally be effective as dispersers, some rarely as establishers, only the nutcracker performs both roles most efficiently.[57] This is because nutcrackers scatter-hoard and cache well beyond their metabolic needs, at depths, to reduce predation and desiccation and in sites favourable for growth, making them potential dispersers capable of regenerating these *Strobus* pine forests.

Based on these scientific insights of similar mutualism between closely affiliated species in temperate montane conifer zones, there is a high chance that bird–pine mutualism in chilgoza pine forests does exist and could play a potential role in promoting natural regeneration. A recent study undertaken to confirm the genetic diversity in chilgoza forests found high rates of cross-fertilisation, indicating that there might be substantial chances of pollen and seed migration from one site to another.[58] The researchers did not provide any explanation for these trends. But I felt that this piece of research evidence could be attributable partially to avian seed dispersal, reinforcing the possibility that nutcracker–pine interaction may be active in Kinnaur. In this eventuality, excessive seed extraction and emerging threats to the chilgoza pine habitat, unravelled in the course of this study, could well jeopardise the survival of the bird and the vital role it may be playing in re-wilding the degrading landscape by promoting natural regeneration.

Reflections on the Environment–Development Discourse

Inferences from Chilgoza Pine

The chilgoza pine case study firmly exemplifies that non-timber forest product (NTFP)[59] extraction is not necessarily a benign way of using forests as envisaged by a surge of pioneering forest policy literature pertaining to the late 1980s.[60] These experts envisioned a strong case for the role NTFPs could play in alleviating poverty, while at the same time checking the rates of tropical deforestation, which was becoming a critical concern. They presumed that because the local population had a stake in the NTFP for economic gains, they would tend to protect the forest resource better. Therefore, permitting forest access for NTFPs was vehemently prescribed as a preferable 'liberal' conservation strategy for ensuring livelihood improvements without compromising the environment, in place of the existing restrictive 'protectionist' conservation paradigms that were not producing the expected beneficial outcomes.

More recent studies have questioned the inevitability of joint positive outcomes for sustainable livelihood and conservation associated with NTFP extraction.[61] This growing body of research work, based on a number of instances of NTFPs and trade, seems to now indicate that while livelihood improvement is certain in most cases, evidence of adverse ecological impacts are becoming increasingly common. As per their research inferences, this could result in long-term deleterious effects on reproduction and regeneration of populations being harvested, demographic structure, genetic patterns of plant populations at the community level, as well as overall ecosystem processes. Lamenting on the staggering extraction levels in the contemporary context, they conclude that regeneration rates can hardly keep pace with the rapid rate of competitive market-oriented harvesting.

The degrading condition of the existing chilgoza pine forest stock manifested in over-lopped trees and the abysmal status of natural regeneration reinforces these vital empirical findings. The emerging scenario also imposes threats to the endemic greater spotted nutcracker that could have sustained the tree species. Increasing commercialisation pressures and reckless harvesting of cones for pine nuts have made livelihood gains and resource conservation exigencies virtually irreconcilable. There is an ardent need to initiate conservation measures for reversing these trends in order to revive the condition of these endangered forests. The pitiable status of a host of other Himalayan NTFPs extracted from the wild that suffer from similar impending threats, such as the Himalayan trillium (*Trillidium govanianum*), the caterpillar fungus (*Ophiocordyceps sinensis*) and the Himalayan yew (*Taxus baccata*), to name a few, raise major biodiversity concerns for these fragile mountain ecosystems.

Just like the presumed inevitability of positive outcomes associated with NTFP extraction, poverty of forest fringe communities is invariably rooted as the principal cause of environmental degradation in the arena of environment–development discourses.[62] But revelations emerging from the chilgoza pine case study, set in an unusually well-endowed tribal belt, are at odds with these diktats. High levels of economic prosperity appeared to be consistent with mounting threats to the environment.

At first glance, as I scrutinised the field survey data more closely, it was unclear to me what propels this colossal scale of resource exploitation. As per the household survey data, returns from pine nut sale account for a mere 1–12 per cent of a household's total annual income. A comparative assessment of income receipts from the sale of chilgoza pine nuts and the sale of apples revealed an anomalous scenario. The mean annual income from pine nuts turned out to be around Rs 7,684 for a household. I inferred that this was a pittance in comparison to a household's mean income from horticulture, which was 30 times higher, amounting to Rs 2.6 lakhs per annum (Figure 7.8). Therefore, despite the relatively meagre significance of pine nuts as a source of livelihood, there is near-total extraction of pine cones for the market. Since poverty is not an issue in this tribal belt and there are alternative income sources that yield many times more, why do people extract? What is the incentive to collect?

Figure 7.8: Comparative Statistics: Household Income from Apple and Pine Nut Sales

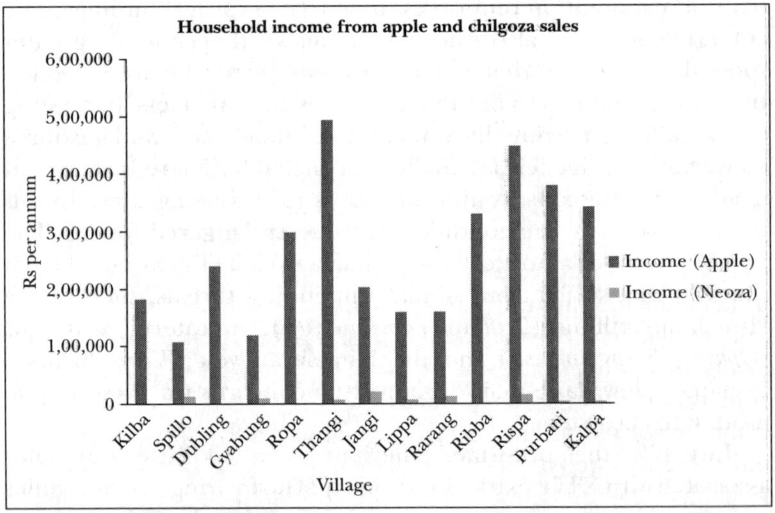

Source: Field studies 2009.

Field interactions initiated to unravel this paradoxical situation exposed some interesting trends. When respondents were queried on what drives extraction in the present context, they were unanimous in conveying that the returns accruing to them were relatively high, commensurate with the time and effort involved in cone collection. Some repented that there was uncertainty about future returns, particularly due to declining yield. So they preferred to exploit as much as possible in the present. Others appeared to be pragmatic while affirming that withdrawing from these engagements altogether they may impinge on their customary rights of access to the natural resource. It was therefore essential to continue resource extraction to retain their rights. The least benefits seemed to accrue to villages that prefer to auction out their forests to *thekedars*. But they continue to initiate these endeavours as they are custom-bound to fulfil their traditional religious obligations towards replenishing the local deity's coffers through auction receipts. These observations do seem to fall in place with Balland and Platteau's[63] findings based on extensive case studies of common property resources across the globe. Their interpretive insights emanating from these studies reveal that market penetration leads to resource depletion because there tends to be considerable uncertainty in the minds of the people as regards future price or resource regulation regime. Thus, they prefer to exploit present opportunities to the fullest possible extent for maximum possible gains.

The Conservation Environment

Evidence on the degenerating chilgoza pine resource condition implies serious lacunae in the existing conservation environment. The inability of the State and the local community to upscale recruitment or promote sustainable management of these endangered temperate pine forests was quite evident from this field endeavour. Ever since the inception of formal state forest governance, this conifer species has been somewhat neglected due to its poor timber value.[64] Qualms about the State's engagement in the upkeep of India's forests are being vociferously expressed,[65] and these are well-justified when viewed from the perspective of the chilgoza forests in Kinnaur.

This chilgoza pine study, which was initiated under the aegis of the State Forest Department, had great potential for a much-needed bridging of the gap between research, policy and action. On completion of my research project, the findings from this study were well-received, drawing attention to a depleting endemic resource. Eventually, these efforts resulted in a well-endowed state action plan for conservation action. However, the abrupt termination of forest officials responsible for initiating the conservation programme came as a blow to the project. The ousted officials were replaced by a new regime that was sceptical about the engagement of external expertise and more inclined towards postponement of necessary action. These narratives aptly expose the volatile nature of the state conservation domain and its closed-mindedness towards credible scientific enquiry, which is a prerequisite to foster effective conservation outcomes.

At the local community level, the alarming status of chilgoza forests adequately substantiates that indigenous governance mechanisms of common resource management may not necessarily target conservation as the prime objective while regulating resource use. Rather, in the chilgoza pine domain, local norms have been judiciously framed to assure that equity prevails during allocation of the valuable pine nut resource across households in the village (Image 7.5). These findings are very much in sync with Balland and Platteau's comprehensive empirical evidence on the efficacy of local governance across a host of commons, which they critically reviewed and analysed.[66] On the basis of these trends, they confirmed that locals tend to be fairly deft at solving distribution or allocation-related problems, but they may not be conservationists inherently, overtly concerned about deteriorating ecological effects or threats to long-term sustainability of the common resource.

Image 7.5: Distribution of Collectively Extracted Pine Cones in Progress

Source: Field studies 2009.

Therefore, my findings definitely bring to light the apathetic nature of the conservation environment, raising serious biodiversity concerns as these endemic chilgoza forests are an integral part of a fragile and critical mountain ecosystem. Under these circumstances, conservation action is imperative to prevent repercussions for the overall environment of the region.

First and foremost, the ecological values of chilgoza pine forests need to be brought to the forefront through research by unearthing its keystone and foundation species characteristics, its slope-stabilising and soil-anchoring properties, its hydrological functions, and its inherent nurturing abilities for supporting other flora and fauna of the region. Vital inputs emanating from these studies can be of great importance for designing resource management strategies to reinforce and revive these key ecological values. For greater effectiveness, these initiatives would need to mobilise the local community in meaningful ways by creating awareness to make them more ecologically responsible towards the degrading resource. A citizen science[67] kind of participatory framework that incorporates a mix of indigenous knowledge and scientific spirit is specifically proposed.[68] It is hoped that this would ensure more resilient socio-ecological systems in future. At the outset, innovative resource-saving technological interventions, such as a sustainable extraction tool, can easily be designed and disseminated. Besides, well-structured incentives can be proposed for rewarding sustainable practices. A few villages can be initially adopted to test these initiatives after assessing the village community's eagerness to participate in such conservation missions. Above all, a synergy between research, policy and action is much warranted as a necessary prerequisite for achieving these conservation goals in addition to promoting long-term monitoring of the resource condition.

Concluding Remarks

My research journey ended on a somewhat disturbing note. Even if my initial ocular impressions of the flurry of cone collection made poor natural regeneration a tenable outcome, I did not anticipate the extent of resource impacts until my field research studies and empirical results unearthed the intensity of the problem. The degrading status of chilgoza pine forests in Kinnaur that is intensifying due to laxity in confronting the poor resource condition at all levels of forest governance, calls for immediate attention to prevent the endemic species from aging away besides mitigating threats to the fragile mountain ecosystem it supports. Emerging scientific evidence on the vulnerability of the Himalayas to climate change[69] and the envisaged primary extinction threats for range-restricted species that have tight environmental requirements, such as the chilgoza pine, raises grave concerns about the sustainability of these rare forests. Based on ground truth evidence emerging from this study and possibly similar insights on other native habitats in Afghanistan and Pakistan, re-categorising the status of this species from 'near threatened' to the International Union for Conservation of Nature (IUCN) red-list of 'endangered' species is deeply warranted for greater awareness and undeterred conservation focus.

By unearthing the changing dynamics of a fragile mountain landscape, the trajectory of my field research endeavour terminated by expounding incessant threats to a rare and endemic species of pine, and also concomitant threats to an endemic bird that is presumed to be an efficient seed disperser for sustaining the forest stock. As artificial propagation strategies have not produced the expected positive results so far, and a concerted conservation drive at all levels is still warranted, managing forests to enhance greater nutcracker visitations by incentivising locals to leave a few cones behind on trees, or by curtailing cone collection and grazing from a few areas on a rotational basis, can be a quick cost-effective strategy for restoring depleting chilgoza pine forests.

Notes

1. 'Wildlife Conservation in India: From Policy to Practice', Based on a Paper Presented at a workshop, Under the Aegis of The Centre for Advanced Studies of India, University of Pennsylvania, 4–5 November 2016, New Delhi.
2. R. Wade, 1987, 'The Management of Common Property Resources: Finding a Cooperative Solution', *Research Observer* 2(2): 219–34.
3. A. Gerard, 1841, *Account of Kunawar in the Himalaya*, G. Lloyd (ed.), 1792–1839 (London: James Madden and Co.).
4. For the household surveys, the sample of households in each

village was carefully selected to replicate the joint distribution of actual landholding and caste structure so that all factions and social groups were represented. The surveys focused on documenting a multiplicity of household characteristics such as assets, income, returns from the sale of pine nuts, the nature of tools used for pine nut collection, and institutional mechanisms for collection of cones at the village level.

5. The relevant forest stands representing pine nut access zones were meticulously mapped for every village by referring to compartment history files maintained by the State Forest Department coupled with added inputs provided by the field-level forest department staff as well as the local community. Once these zones were identified, circular plots of size 100 sq. m were randomly laid to ensure that these chilgoza pine belts were extensively covered for gauging a host of forest attributes using conventional forestry science tools.
6. D. M. Richardson and P. W. Rundel, 1998, 'Ecology and Biogeography of Pinus: An Intoduction', in C. M. Richardson (ed.), *Ecology and Biogeography of Pinus*, pp. 3–46 (Cambridge, UK: Cambridge University Press).
7. More specifically, the species is categorised under the subgenus *Strobus*; section *Quinquefoliae*; and subsection *Gerardianae.*
8. The two other East Asian species in the *Gerardianae* subsection are *Pinus bungeana* and *Pinus squamata.* Studies based on genetic traits have found *Pinus squamata* to be the closest associate of *Pinus gerardiana* (Zhang et al. 2003; Hao et al. 2015). In terms of species abundance, the two affiliated tree species are quite apart. *Pinus bungeana* or lace bark pine is widely cultivated. However, *Pinus squamata* is the most critically endangered pine species in the world. A single locality in the mountainous Yunan province of China hosts 20 trees, the only living global specimens of this near-extinct taxon.
9. R. S. Troup, 1921, *Silviculture of Indian Trees: Lauraceae to Coniferae*, Vol. III, pp. 1090–93 (Oxford: Clarendon Press); Thomson 1852.
10. C. I. Millar, 1993, 'Impact of Eocene on the Evolution of Pinus', *Annual Missouri Botanical Garden* 80(2): pp. 471–98.
11. Biogeography refers to the geographical occurrence of a species and its related genetic structure.
12. Z. Hao, Y. Liu, M. Nazare, et al., 2015, 'Molecular Phylogenetics and Evolutionary History of Sect. Quinquefoliae (Pinus): Implications for Northern Hemisphere Biogeography', *Molecular Phylogenetics and Evolution* 87: 65–79.
13. Z. Hao et al., 'Molecular Phylogenetics'; Liu et al., 'Phylogeny of

Pinus armandii'; Zhang et al., 'Phylogenetic Relationship of an Extremely Endangered Species'.

14. E. O. Wilson, 1988, 'The Current State of Biological Diversity', in E. O. Wilson, and F. M. Peters (eds), *Biodiversity*, pp. 3–18, (Washington, D.C.: The National Academies of Sciences, Engineering, and Medicine).
15. M. Ahmed, M. Ashfaq, M. Amjad, et al., 1991, 'Vegetation Structure and Dynamics of *Pinus gerardiana* Forests in Balouchistan, Pakistan', *Journal of Vegetation Science* 2(1): 119–24.
16. J. Singh and R. Yadav, 2007, 'Dendroclimatic Potential of Millenium Long Ring Width Chronology of *Pinus gerardiana* from Himachal Pradesh, India', *Current Science* 93(6): 833–36.
17. Troup, *Silviculture of Indian Trees*, pp. 1090–93.
18. M. Barbero, R. Loisel, And P. Quezel, 1992, 'Biogeography, Ecology and History of Mediterranean *Quercus ilex Ecosystems*', in F. Romane and J. Terradas (eds), *Quercus ilex L. Ecosystems: Functions, Dynamics and Management* Series, Vol. 13, *Advances in Vegetation Science*, pp. 19–34 (Netherlands: Springer).
19. S. H. Champion and S. K. Seth, 1968, *A Revised Survey of the Forest Types of India*, p. 6 (New Delhi: Manager of Publications, Government of India).
20. Thomson, *Western Himalaya and Tibet*; Gerard, *Áccount of Kunawar in the Himalaya.*
21. R. M. Lanner, 1998, 'Seed Dispersal in Pinus', in D. M. Richardson (ed.), *Ecology and Biogeography of Pinus*, pp. 281–95 (Cambridge, London: Cambridge University Press).
22. D. F. Tomback and Y. B. Linhart, 1990, 'The Evolution of Bird-dispersed Pines', *Evolutionary Ecology* 4(3): 185–219.
23. Gerard, *Áccount of Kunawar in the Himalaya*; Thompson, *Western Himalaya and Tibet*; R. Sarkar, *Unpublished Field Notes and Oral History accounts of District Kinnaur.*
24. Troup, *Silviculture of Indian Trees*, pp. 1090–93.
25. H. Kapadia, 1987, 'A Note on Kinnaur', *Himalayan Journal* 43(14): 30.
26. Gerard, *Áccount of Kunawar in the Himalaya.*
27. M. Kenneth, 1936, 'Himalaya as a Barrier to Modern Communication', *The Geographical Journal* 87(1): 1–13.
28. Gerard, *Áccount of Kunawar in the Himalaya*; Sarkar, *Unpublished Field Notes.*
29. R. Sarkar, 2008, *Why Diverse Common Property Access Regimes Co-exist and Transform. Evidence from an Ecological Fragile Belt of the Himalayas*, available at https://dlc.dlib.indiana.edu/dlc/handle/10535/742.

30. Troup, *Silviculture of Indian Trees*, pp. 1090–93.
31. Gerard, *Áccount of Kunawar in the Himalaya.*
32. Troup, *Silviculture of Indian Trees*, pp. 1090–93.
33. R. Sarkar, 2007, Interview with R. S. Negi, Environmentalist, Kinnaur on the 'Vexed Question of Transformation and Change in Fragile Mountain Areas', Seminar, Vol. 577, October, available at http://www.india-seminar.com/semframe.html.
34. A. Sharma, 1999, *An American in Khadi: The Definitive Biography of Satyanand Stokes*, p. 109, 113, 226–35 (USA: Penguin Books).
35. R. Sarkar, 2010, 'Rural Accessibility and Development: Sustainability Concerns in an Ecologically Fragile Mountain Belt', *Economic and Political Weekly* XLV(21): 63–71.
36. P. D. Sharma and R. S. Minhas, 1993, 'Land Use and Bio-Physical Environment of Kinnaur District, Himachal Pradesh, India', *Mountain Research and Development* 13(1): 41–60.
37. ibid.
38. Source: District Statistical Office, Recong Peo, Kinnaur.
39. R. Sarkar, 'Rural Accessibility and Development'.
40. Gerard, *Áccount of Kunawar in the Himalaya.*
41. *Chilgoza* pine nuts, along with those of *Pinus pinea* and *Pinus koraiensis*, are among the most widely used and exported. The long and tapering shape of these pine seeds are preferred by confectionery industries as they are easier to insert into cakes and sweets than the blunt nuts of most other edible pinõn pines (Ciesla 2001).
42. Pine nuts extracted from Afghanistan also enter retail markets in the urban centres of Northern India. But there are no reliable estimates of the total magnitude originating from this source.
43. Troup, *Silviculture of Indian Trees*, pp. 1090–93.
44. ibid.
45. Local estimates revealed that nearly 100–300 metric tonnes of pine nuts are exported out of Kinnaur each year for sale.
46. Frequency distribution depicted in Figure 7.4 is based on a forest assessment initiated in 327 circular plots of 100 sq. m each across the chilgoza forests in the surrounding vicinity of the case study villages accessed by local inhabitants.
47. D. M. Bhat and N. H. Ravindranath, 1995, 'Effect of Lopping Intensity on Tree Growth and Stand Productivity in Tropical Forests', *Journal of Tropical Forest Science* 8(1): 15–23; S. Singh, 2002, 'Balancing the Approaches of Environmental Conservation by Considering Ecosystem Services as well as Biodiversity', *Current Science* 73(4): 371–74; P. Wohlleben, 2016, *The Hidden Life of Trees.*

What they Feel, how they Communicate. Discoveries from a Secret World (Gurgaon, Haryana: Penguin Books).

48. A mast season is one when there is a boom in seed production, which is a natural feature of the pine species.
49. G. Shahabuddin and S. Prasad. 2004. 'Assessing Ecological Sustainability of Non-Timber Forest Produce Extraction: The Indian Scenario', *Conservation and Society* 2(2): 236–48.
50. Troup, *Silviculture of Indian* Trees; R. N. Sehgal and P. K. Sharma, 1989, 'Chilgoza the Endangered Social Forestry Pine of Kinnaur', Technical Bulletin No. FBTI, 1 (Nauni, Solan: Dr. Y.S. Parmar University of Horticulture and Forestry); D. M. Richardson and P. W. Rundel, 'Ecology and Biogeography of Pinus'; R. Peltier and V. Dauffy, 2009, 'The Chilgoza of Kinnaur. Influence of the *Pinus gerardiana* Edible Seed Market Chain Organisation on Forest Regeneration in the Indian Himalayas', *Fruits* 54(2): 91–110.
51. Troup, *Silviculture of Indian Trees,* pp. 1090–93.
52. H. E. Hutchins and R. M. Lanner, 1982, 'The Central Role of Clark's Nutcracker in the Dispersal & Establishment of Whitebark Pine', *Oecologia* 55(2): 192–201.
53. Tomback and Linhart, 'The Evolution of Bird-Dispersed Pines'.
54. Lanner, *Seed Dispersal in Pinus,* pp. 281–95.
55. M. B. Pesendorfer, T. S. Sillet, W. D. Koenig, et al., 2016, 'Scatter-hoarding Corvids as Seed Dispersers for Oaks and Pines: A Review of a Widely Distributed Mutualism and its Utility to Habitat Restoration', *The Condor: Ornithological Applications* 118(2): 215–37.
56. H. E. Hutchins and R. M. Lanner, 1982, 'The Central Role of Clark's Nutcracker'; H. E. Hutchins, S. A. Hutchins, and B. Liu, 1996, 'The Role of Birds and Mammals in Korean pine (*Pinus koraiensis*) Regeneration Dynamics', *Oecologia,* 107(1): 120–30; M. Hayashida, 2003, 'Seed Dispersal of Japanese Stone Pine by the Eurasian Nutcracker', *Ornithological Science,* Ornithological Society of Japan 2(1): 33–40.
57. Hutchins and Lanner, 'The Central Role of Clark's Nutcracker'.
58. A. Kant, D. Pattanayak, S. K. Chakrabarti, et al., 2006, 'RAPD Analysis of Genetic Variability in *Pinus gerardiana* in Kinnaur, Himachal Pradesh', *Indian Journal of Biotechnology* 5(1): 62–67.
59. NTFPs are non-wood animal and plant parts, such as chilgoza seeds, extracted from the wild.
60. N. Myres, 1988, 'Tropical Forests: Much More Than Stocks of Wood', *Journal of Tropical Ecology* 4(2): 209–21; C. M. Peters, A. H. Gentry, and R. O. Mendelsohn, 1989, 'Valuation of an Amazonian Rain Forest', *Nature* 339: 655–56.

61. J. E. M. Arnold and M. R. Perez, 2001, 'Can Non-Timber Forest Products Match Tropical Forest Conservation and Development Objectives?', *Ecological Economics* 39(3): 437–47; Shahabuddin and Prasad, 'Assessing Ecological Sustainability'; K. Kusters, R. Achdiawan, B. Belcher, et al., 2006, 'Balancing Development and Conservation? An Assessment of Livelihood and Environmental Outcomes of Nontimber Forest Product Trade in Asia, Africa and Latin America', *Ecology and Society* 11(2): 20; S. Gubbi and C. MacMillan, 2007, 'Can Non-Timber Forest Products Solve Livelihood Problems? A Case Study from Periyar Tiger Reserve, India', *Fauna and Flora International, Oryx* 42(2): 222–28; M. Z. Steele, C. M. Shackleton, R. U. Shanker, et al., 2015, 'The Influence of Livelihood Dependency, Local Ecological Knowledge and Market Proximity on Ecological Impacts of Harvesting NTFPs', *Forest Policy and Economics,* 50(C): 285–91.
62. A. K. Duraiappah, 1998, 'Poverty and Environmental Degradation: A Review and Analysis of Nexus', *World Development* 26(12): 2169–79; A. Angelson and S. Wunder, 2003, 'Exploring the Forest-Poverty Link: Key Concepts, Issues and Research Implications', CIFOR Occasional Paper, No. 40, Centre for International Forestry Research, Bogor, Indonesia.
63. J. M. Balland and J. P. Platteau, 2000, *Halting Degradation of Natural Resources: Is there a Role for Rural Communities,* p. 273 (Oxford: FAO, United Nations and Oxford University Press).
64. Troup, *Silviculture of Indian Trees,* pp. 1090–93.
65. K. Kohli and M. Menon, 2015, 'Growing Forests in the Air', *The Hindu, Future Forecast,* 26 October 2015.
66. Balland and Platteau, *Halting Degradation of Natural Resources,* p. 273.
67. A. H. Toomey and M. C. Domroese, 2013, 'Can Citizen Science Lead to Positive Conservation Attitudes and Behaviours', *Research in Human Ecology* 20: 50–62.
68. The olive ridley turtle conservation success story (Shankar 2016; Lopez 2016) and the snow leopard conservation success story (Mishra 2016) are motivating examples which have incorporated such approaches.
69. U. B. Shrestha, S. Gautam, and K. Bawa, 2012, 'Widespread Climate Change in the Himalayas and Associated Changes in Local Ecosystems', *PLOS ONE* 7(5)e36741: 1–10; A. El-Keblawy, 2014, 'Impact of Climate Change on Biodiversity Loss and Extinction of Endemic Plants of Arid Land Mountains', *Journal of Biodiversity and Endangered Species* 2(1) 1–3.

REFERENCES

Ahmed, M., M. Ashfaq, M. Amjad, et al. 1991. 'Vegetation Structure and Dynamics of *Pinus gerardiana* Forests in Balochistan, Pakistan', *Journal of Vegetation Science* 2(1): 119–24.

Angelson, A., and S. Wunder. 2003. 'Exploring the Forest-Poverty Link: Key Concepts, Issues and Research Implications', CIFOR Occasional Paper, No. 40, Centre for International Forestry Research, Bogor, Indonesia.

Arnold, J. E. M, and M. R. Perez. 2001. 'Can Non-Timber Forest Products Match Tropical Forest Conservation and Development Objectives?', *Ecological Economics* 39(3): 437–47.

Balland, J. M., and J. P. Platteau. 2000. *Halting Degradation of Natural Resources: Is there a Role for Rural Communities,* p. 273. Oxford: FAO, United Nations and Oxford University Press.

Barbero, M., R. Loisel, and P. Quezel. 1992. 'Biogeography, Ecology and History of Mediterranean *Quercus ilex* Ecosystems'. In F. Romane and J. Terradas (eds), *Quercus ilex L. Ecosystems: Functions, Dynamics and Management* Series, Vol. 13, *Advances in Vegetation Science,* pp. 19–34. Netherlands: Springer.

Bhat, D. M., and N. H. Ravindranath. 1995. 'Effect of Lopping Intensity on Tree Growth and Stand Productivity in Tropical Forests', *Journal of Tropical Forest Science* 8(1): 15–23.

Champion, S. H., and S. K. Seth. 1968. *A Revised Survey of the Forest Types of India,* p. 6. New Delhi: Manager of Publications, Government of India.

Ciesla, W. M. 2001. 'Non-Wood Forest Products from Conifers', pp. 72–85. Rome: Food and Agriculture Organisation of the United Nations (FAO).

District Statistical Office, Recong Peo, Kinnaur.

Duraiappah, A. K. 1998. 'Poverty and Environmental Degradation: A Review and Analysis of Nexus', *World Development* 26(12): 2169–79.

El-Keblawy, A. 2014. 'Impact of Climate Change on Biodiversity Loss and Extinction of Endemic Plants of Arid Land Mountains', *Journal of Biodiversity and Endangered Species* 2(1): 1–3.

Gerard, A. 1841. *Áccount of Kunawar in the Himalaya,* G. Lloyd (ed.), pp. 1792–1839. London: James Madden and Co.

Gubbi, A., and C. MacMillan. 2007. 'Can Non-Timber Forest Products Solve Livelihood Problems? A Case Study from Periyar Tiger Researve, India', *Oryx* 42(2): 222–28.

Hao, Z., Y. Liu, M. Nazare, et al. 2015. 'Molecular Phylogenetics and Evolutionary History of Sect. Quinquefoliae (Pinus): Implications for Northern Hemisphere Biogeography', *Molecular Phylogenetics and Evolution* 87: 65–79.

Hayashida, M. 2003. 'Seed Dispersal of Japanese Stone Pine by the Eurasian Nutcracker', *Ornithological Science* 2(1): 33–40.

Hutchins, H. E., and R. M. Lanner. 1982. 'The Central Role of Clark's Nutcracker in the Dispersal & Establishment of Whitebark Pine', *Oecologia* 55(2): 192–201.

Hutchins, H. E., S. A. Hutchins, and B. Liu. 1996. 'The Role of Birds and Mammals in Korean pine (*Pinus koraiensis*) Regeneration Dynamics', *Oecologia,* 107(1): 120–30.

Kant, A., D. Pattanayak, S. K. Chakrabarti, et al. 2006. 'RAPD Analysis of Genetic Variability in *Pinus gerardiana* in Kinnaur, Himachal Pradesh', *Indian Journal of Biotechnology* 5(1): 62–67.

Kapadia, H. 1987. 'A Note on Kinnaur', *Himalayan Journal* 43(14): 30.

Kenneth, M. 1936. 'Himalaya as a Barrier to Modern Communication', *The Geographical Journal* 87(1): 1–13.

Kohli, K., and M. Menon. 2015. 'Growing Forests in the Air', *The Hindu, Future Forecast,* 26 October 2015.

Kusters, K., R. Achdiawan, B. Belcher, et al. 2006. 'Balancing Development and Conservation? An Assessment of Livelihood and Environmental Outcomes of Nontimber Forest Product Trade in Asia, Africa and Latin America', *Ecology and Society* 11(2): 20.

Lanner, R. M. 1998, 'Seed Dispersal in Pinus'. In D. M. Richardson (ed.), *Ecology and Biogeography of Pinus,* pp. 281–95 (Cambridge, London: Cambridge University Press).

Liu, L., Z. Hao, Y. Liu, et al. 2014. 'Phylogeny of *Pinus armandii* and its Relatives: Heterogenous Contributions of Geography and Climate Changes to Genetic Differentiation and Diversification of Chinese White Pines', *PLOS One* 9(1), e85920: 1–12.

Lopez, R. 2016. 'Meet the Home Grown Guardians of Odisha's Olive Ridley Turtles', *The Hindustan Times,* 20 March 2016.

Millar, C. I. 1993. 'Impact of Eocene on the Evolution of Pinus', *Annual Missouri Botanical Garden* 80(2): pp. 471–98.

Mishra, C. 2016. *The Partners Principle for Community-Based Conservation,* Seattle: Snow Leopard Trust.

Myres, N. 1988, 'Tropical Forests: Much More Than Stocks of Wood', *Journal of Tropical Ecology* 4(2): 209–21.

Peltier, R., and V. Dauffy. 2009. 'The Chilgoza of Kinnaur. Influence of the *Pinus gerardiana* Edible Seed Market Chain Organisation on Forest Regeneration in the Indian Himalayas', *Fruits* 54(2): 91–110.

Pesendorfer, M. B., T. S. Sillet, W. D. Koenig, et al. 2016. 'Scatter-hoarding Corvids as Seed Dispersers for Oaks and Pines: A Review of a Widely Distributed Mutualism and its Utility to Habitat Restoration', *The Condor: Ornithological Applications* 118(2): 215–37.

Peters, C. M., A. H. Gentry, and R. O. Mendelsohn. 1989. 'Valuation of an Amazonian Rain Forest', *Nature* 339: 655–56.

Richardson, D. M., and P. W. Rundel. 1998. 'Ecology and Biogeography of Pinus: An Intoduction'. In C. M. Richardson (ed.), *Ecology and Biogeography of Pinus*, pp. 3–46. Cambridge, UK: Cambridge University Press.

Sarkar, R. 2007. Interview with R. S. Negi, Environmentalist, Kinnaur on the 'Vexed Question of Transformation and Change in Fragile Mountain Areas', Seminar, Vol. 577, October, available at http://www.india-seminar.com/semframe.html.

———. 2008. 'Why Diverse Common Property Access Regimes Co-exist and Transform. Evidence from an Ecological Fragile Belt of the Himalayas', Conference Paper, Indiana University. Available at https://dlc.dlib.indiana.edu/dlc/handle/10535/742.

———. 2009. *Unpublished Field Notes and Oral History accounts of District Kinnaur.*

———. 2010. 'Rural Accessibility and Development: Sustainability Concerns in an Ecologically Fragile Mountain Belt', *Economic and Political Weekly* XLV(21): 63–71.

Sehgal, R. N., and P. K. Sharma. 1989. 'Chilgoza the Endangered Social Forestry Pine of Kinnaur', *Technical Bulletin No. FBTI*, 1. Nauni, Solan: Dr. Y. S. Parmar University of Horticulture and Forestry.

Shahabuddin, G., and S. Prasad. 2004. 'Assessing Ecological Sustainability of Non-Timber Forest Produce Extraction: The Indian Scenario', *Conservation and Society* 2(2): 236–48.

Shankar, K. 2016. 'From Soup to Superstar: The Story of Sea Turtle Conservation along the Indian Coast'. New Delhi: HarperCollins Publishers.

Sharma, A. 1999, *An American in Khadi: The Definitive Biography of Satyanand Stokes*, pp. 109, 113, 226–35. USA: Penguin Books.

Sharma, P. D., and R. S. Minhas. 1993. 'Land Use and Bio-Physical Environment of Kinnaur District, Himachal Pradesh, India', *Mountain Research and Development* 13(1): 41–60.

Shrestha, U. B., S. Gautam, and K. Bawa. 2012. 'Widespread Climate Change in the Himalayas and Associated Changes in Local Ecosystems', *PLOS ONE* 7(5)e36741: 1–10.

Singh, J., and and R. Yadav. 2007. 'Dendroclimatic potential of Millenium Long Ring Width Chronology of *Pinus gerardiana* from Himachal Pradesh, India', *Current Science* 93(6): 833–36.

Singh, S. 2002. 'Balancing the Approaches of Environmental Conservation by Considering Ecosystem Services as well as Biodiversity', *Current Science* 73(4): 371–74.

Steele, M. Z., C. M. Shakleton, R. U. Shanker, et al. 2015. 'The Influence of Livelihood Dependency, Local Ecological Knowledge and Market Proximity on Ecological Impacts of Harvesting NTFPs', *Forest Policy & Economics*, 50(C): 285–91.

Thomson, T. 1852. *Western Himalaya and Tibet: A Narrative of a Journey through the Mountains of N. India During the Years 1847–1848*, pp. 63–97. London: Reeve and Co.

Tomback, D. F., and Y. B. Linhart. 1990. 'The Evolution of Bird-dispersed Pines', *Evolutionary Ecology* 4(3): 185–219.

Toomey, A. H., and M. C. Domroese. 2013. 'Can Citizen Science lead to Positive Conservation Attitudes and Behaviours', *Research in Human Ecology* 20: 50–62.

Troup, R. S. 1921. *Silviculture of Indian Trees: Lauraceae to Coniferae*, Vol. III, pp. 1090–93. Oxford: Clarendon Press.

Wade, R. 1987. 'The Management of Common Property Resources: Finding a Cooperative Solution', *Research Observer* 2(2): 219–34.

'Wildlife Conservation in India: From Policy to Practice', Paper Presented at a Workshop, Under the Aegis of The Centre for Advanced Studies of India, University of Pennsylvania, 4–5 November 2016, New Delhi.

Wilson, E. O. 1988. 'The Current State of Biological Diversity'. In E. O. Wilson, and F. M. Peters (eds), *Biodiversity*, pp. 3–18. Washington, D.C.: The National Academies of Sciences, Engineering, and Medicine.

Wohlleben, P. 2016. *The Hidden Life of Trees: What they Feel, how they Communicate. Discoveries from a Secret World.* Gurgaon, Haryana: Penguin Books.

Zhang, Z. Y., J. B. Yang, and D. Z. Li. 2003. 'Phylogenetic Relationship of an Extremely Endangered Species, *Pinus sqamata* (Pinaceae) Inferred from Four Sequences of the Chloroplast Genome and ITS of the Nuclear Ribosomal DNA', *Acta Botanica Sinica* 45(5): 530–35.

8

Conservation in Times of Development

Forest Change Trajectories in State and Community Forests of Jaintia Hills, Meghalaya

RAJKAMAL GOSWAMI AND T. GANESH

On 26 May 2017, the Prime Minister of India, Narendra Modi, inaugurated the country's longest 'Dhola-Sadiya' bridge, measuring 9.15 kilometres in length, over river Lohit in Assam. While dedicating the bridge to the nation, he proclaimed, 'The bridge will not only save money, reduce travel time but it comes as a foundation for the beginning of a new economic revolution. For fulfilling India's dream of becoming a top economically developed nation, a superpower, then [sic] the North Eastern region is the place to begin with.'[1]

The statement of the Prime Minister was in tune with the growth and development-oriented outlook of the Indian polity since the economic reforms of 1991. But, it was perhaps for the first time in the history of Independent India that its Prime Minister had proclaimed the north-eastern region to play a central role in India's journey towards fulfilling its ordained destiny of becoming an economic superpower. This new role is in deep contradiction to the conventional perception of the northeast[2] as India's most backward and underdeveloped zone nestled amidst dense mountainous jungles, where primitive 'tribal' groups are perennially engaged in ethnic conflicts and insurgency.

Such an image has discouraged industrialists and entrepreneurs to invest in the region, while many existing operations in tea and oil industries were either abandoned or very badly affected by insurgency-related extortion demands and ethnic violence. Due to such disturbances, the region was considered unconducive for sustained economic activity in spite of its vast reserves of oil, mineral and natural resources, and its geopolitically and strategically important location (Hazarika 1994). The region's close proximity to Southeast Asia, with which the northeast was connected not only by contiguity of land and culture, but also by historical trade and markets, was seen as having a huge but untapped economic potential. Several massive infrastructure and industrial development projects were planned to exploit these trade linkages, but they failed to take off due to persistent ethnic conflicts and insurgency (Anonymous 2008). Beyond these, the inherent minimum interference policies of the colonial administration and its adoption by the subsequent governments of independent India meant that the northeast remained marginal to the national economic priorities of the 'mainland'.[3] The region was, in Pradip Phanjoubam's evocative description, 'India's area of darkness' wherein 'reluctance to look East, inherent in the overall Indian character, should explain to a good extent why the North East has remained India's area of darkness for so long.'[4]

However, soon insurgency itself became an argument for the Indian government to expand the economic and industrial growth to the frontier hill regions of the northeast. The geopolitical and strategic importance of the region, 98 per cent of whose boundary is shared with Bangladesh, Burma, China and Bhutan, and where most insurgents operated from, further necessitated the upgrade of its economy, military and infrastructure, particularly after the Sino-Indian War of 1962. Thus, militarisation of the region and diplomacy became essential tools to not only counter insurgency and ethnic violence, secure the region's borders from the expansionist ambition of neighbours such as China,[5] and control cross-border insurgency, but to also lay the foundation of 'peace', upon which industries, infrastructure and utilisation of natural and mineral resources could thrive (Baruah 2003). Wherever it helped to contain violent movements, state boundaries were reorganised to create new ethno-political entities,[6] while in others, intensive armed operations, buffered by laws which provided immunity to the army against the military excesses, were used.[7] Apart from the overt efforts of the Indian State, the reduction in tensions around narrow ideas of ethnicity and regionalism were facilitated by the 1991 reforms as they created greater opportunities for trade and cultural exchanges due to the movement of goods and people (Papola 2017). The subsequent revolutionary overhaul of information and communication technology further helped to blunt the effects of such a divisive and separatist rhetoric on popular imagination.

The considerable ease in the ethnic and insurgency-based violence by the late 1990s, along with several policy-driven incentives around the same time to aid industrialisation and exploitation of yet untapped natural, mineral and nuclear resources, spurred economic development activities in large parts of the northeast. During this time, the per-capita net state domestic product (per-capita NSDP) of the region doubled from Rs 13,633 in 1991–92 to Rs 27,896 in 2013–14 at 2004–05 prices with many states such as Mizoram, Nagaland and Tripura almost quadrupling their per-capita NSDP during the same period.[8]

Such rapid growth, however, has come at a substantial cost to the natural environment, and the traditional socio-political fabric and institutions of the region (Goswami et al. 2012; Shrivastava 2015). It led to the flouting of laws and regulations along with the erosion of the traditional rights and access of the communities over their resources, such as forests, water and minerals, while also endangering their livelihood options (Sharma and Shrivastava 2011). In many cases, large-scale industrialisation, damming and unregulated mining became flashpoints of public resistance owing to their real and potential impacts on rivers, forests and people (Duarah 2017; Goswami and Jesudasan 2012; Kashyap 2016).

And yet, there are few empirical studies which have tried to understand the impacts of such large-scale developmental activities on the north-eastern region's forest resources and biodiversity since 1990. This is particularly important, given the unique positioning of the region, where greater than 85 per cent of the forests are under the control of the communities. While state-managed forests are negligible in terms of area, they are important reservoirs of wildlife and their habitat (Goswami 2015; Velho et al. 2016). Therefore, it is important to understand how the two dominant types of land and forest ownership/management regimes fare against the onslaught of infrastructure and industrial development, which is expected to further intensify in the coming years.

There are two broad scales on which the success or failure of forest conservation can be evaluated. One is to look at the temporal scale within the same landscape(s), which typically requires collection of data over long periods of time to come up with any meaningful conclusions, given the relatively slower speeds at which biological processes express themselves. The second one is to compare reliable and relevant outcomes across two or more management systems of interest within a relatively short period of time, in a way, a 'snapshot study'. Therefore, if the question relates to the understanding of how a difference in the management regime affects forest conservation within the same landscape, given the same socio-political factors, looking at the time-series data is more meaningful.

Collecting data over large temporal scales have been made easy due to rapid developments in the Remote Sensing (RS) and Geographical Information System (GIS) technology, making it possible to monitor land use and land cover (LULC) changes across time and space at fine spatial scales (Jensen 2000; Munyati and Kabanda 2009). Additionally, they allow us the flexibility of designing and performing retrospective analysis of forest cover using multiple time-periods across socio-political axes such as conservation approaches or land-management systems (Lambin and Meyfroidt 2010). Since the absence of natural habitat automatically endangers a high proportion of native biodiversity in the tropical forests (Terborgh and van Schaik 2002), the ability to maintain habitat and forest cover over time may be employed as an appropriate proxy to assess conservation effectiveness (Parrish et al. 2003).

Using the case of the Jaintia Hills in Meghalaya in northeast India, we analyse the patterns of forest cover change across a state-controlled reserve forest (RF) and a community forest (CF) over two decades, with the backdrop of the economic context of fast-paced infrastructural and industrial development in the north-eastern state. We also take a closer look at the drivers of forest cover change in both the state-governed and community-controlled areas, and compare the processes involved. Finally, we discuss how economic policies shape distant landscapes and how different management regimes respond to such drivers in terms of forest change.

Jaintia Hills in Meghalaya

Meghalaya is one of the newer states of India, attaining statehood in 1972 following the reorganisation of undivided Assam. With an area of 22,429 sq. km and a population of approximately 29.5 lakh as per the 2011 Census, Meghalaya has high forest cover (greater than 72 per cent of the total land area) (Forest Survey of India [FSI] 2013), which exhibits considerable overlap with valuable minerals like coal and limestone. Eighty-nine per cent of these forests are under the direct or *de facto* control of the communities, with the Meghalaya Forest Department (FD hereafter) controlling the remaining 11 per cent. Out of these, only 3 per cent are protected as national parks and wildlife sanctuaries, with the rest being reserved and protected forests.

Among the CFs in Meghalaya, there are sacred forests, protected largely due to their religious values, which are scattered and small in size (the largest one measuring 200 hectares) (Tiwari et al. 1998). Further, large-scale conversions of the local communities to Christianity have eroded the sacred values associated with such forests, as a number of studies conducted in these forests have shown. In addition, however, there are large areas of dense CF areas, particularly along the southern slopes of the Khasi-Jaintia plateau and Garo Hills, the status of which remains largely unknown. It is important to monitor these forests as many of these occur in mineral-rich areas, for which there has been a growing demand since the 1990s.

The Hills and the Plateau

There are two distinct types of landscapes in Meghalaya: 1) the centrally situated 'plateau', and 2) the 'hills', which include the southern precipitous zone, and the northern undulating and western hills. The dynamics and scale of resource extraction, livelihood patterns, and economic and forestry activities are vastly different between the plateau area and the steep hills. The differences emerged chiefly due to the local knowledge systems that the communities generated of these landscapes over long periods. Such knowledge drove their respective management and cultural beliefs, which ultimately reshaped those landscapes to their current state (Allen 1905). Figure 8.1 shows the how the distribution of landscape features such as elevation, slope and crop patterns are influenced by the geography of the Meghalaya plateau.

Figure 8.1: Distribution of Cash-crops Along Altitudinal Gradient and North–South Axis in Meghalaya

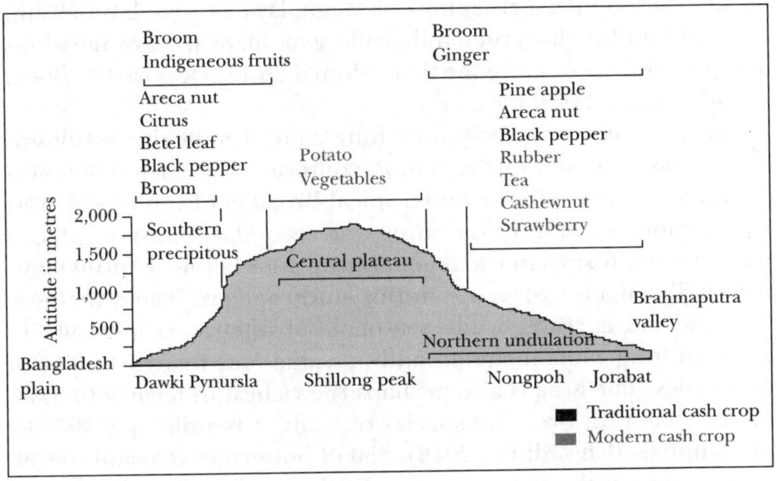

Note: The distribution of cash crops is shown over the major hypsographic regions: the northern undulation, the central plateau and the southern precipitous zone; not to scale.
Source: Behera et al. 2016.

The elevation of the plateau ranges from 900–1600 metres in most places with the highest point recorded in the Shillong Peak at 1,961 metres. The vegetation comprises the Khasi pine (*Pinus kesiya*)-dominated stands, which were believed to be the result of aggressive plantation for timber extraction during the colonial times (Saikia 2011). The plateau area is the most densely populated region and represents the social, political and cultural mainstream of Meghalaya.

Most of the coal occurs in this belt, mining for which is mostly carried out in the sub-terrain using a rudimentary technique called 'rat-hole' mining. The world-famous high precipitation zones, too, are situated within what is known as the Khasi plateau. Owing to a long history of commercial pine plantation, mining and intensive grazing, the natural vegetation of this belt, classified as Sub-tropical

Broadleaf Evergreen Forest by Champion and Seth (1968), is now restricted to small patches, which are either situated in steep, inhospitable slopes or are sacred forests (Tiwari et al. 1998). Between the pine and the sacred groves lie the rolling grasslands which are managed by annual burning by the local communities, the grass being used as fodder for the farm cattle. This landscape evokes the popular perception of Meghalaya as a cool, mountain state with undulating green hills, rolling highlands, grassy meadows and pine forests, leading the first colonial chroniclers to describe it as the 'Scotland of the East'.[9]

In the southern precipitous hilly zone, where this study was carried out, the natural vegetation comprises the Cachar Tropical Evergreen Forest and the Sub-tropical Broadleaf Evergreen Forest (Champion and Seth 1968), along with secondary moist-deciduous and bamboo forests in the regenerating *jhum* fallows (Anonymous 1995). The main cash crops in this landscape are betel nut, betel leaf and citrus fruits, while seasonal subsistence crops sown in small shifting plots include paddy, sesame and many varieties of vegetables. The area is also perhaps the richest in terms of faunal biodiversity with over 600 species of birds (Choudhury 2014), 65 of mammals (Choudhury 2013), 450 of butterflies (Goswami et al. 2015), along with several species of fish, amphibians and reptiles. Limestone is a major mineral mined in this landscape with about an estimated reserve of 15,100 million tonnes,[10] of which a substantial proportion comprises 'cement grade' that has a high demand in the cement industry.

Management of Forests

Constitutionally, the CFs of Meghalaya fall within the three-tier administrative structure due to the presence of the intermediary institution of the Autonomous District Councils (ADCs). While the FD remains responsible for enacting and enforcing laws and policies pertaining to forest management, the ADCs, as will be evident in the subsequent section, are entrusted to frame and implement their own management and conservation-related rules in the community-owned and managed areas. On the ground however, the *dolois* (a traditional chief), who presides over an *elaka* (usually a cluster of 10–40 villages) is the most important and influential institution with respect to land and forest governance in the Jaintia Hills. They function under the Jaintia Hills ADC or the Jaintia Hills Autonomous District Council (JHADC). The original clans, who first formed/founded their respective *elakas* dating back to 500–800 years ago, are very important, and only their members (males alone) can stake a claim to the seat of the *doloi* (Gassah 1998). At the village level, the presiding authority for framing and implementing rules around forest management and resource extraction is the village headman (*waheh chnongs*), who, like the *doloi,* is always a male. However, unlike the *doloi,* the headman is either selected unanimously by the male elders and community members, or through a voting process in which only the adult male members of the village can participate.

The state forests are managed by the FD, whose structure and function resemble those of any other forest department of the country. On ground, however, the situation and context differ even when compared to another state of the northeast region, Assam, where its FD is an all-powerful department when it pertains to the forest and wildlife of the state. In Meghalaya, however, owing to the miniscule proportion of the forests it controls (10.4 per cent), the department is somewhat disempowered when it comes to the management and protection of forests and wildlife beyond the forests of its own jurisdictions. A substantial proportion of the forests (89.6 per cent) is directly under the control of those individuals or community who own it—*de jure* or *de facto.*

Our land use change analysis was carried out across these two broad management and ownership regimes (RF and CF) that we used as a framework for the study. We compared the LULC changes over 20 years across a 231.94 sq. km community-managed Narpuh *elaka* (representing the CF) and a 169.64 sq. km state-managed reserve forest (representing the RF). Both were located in the southern hill slopes of the East Jaintia Hills district.

While the sample size is limited, we believe that an in-depth examination of forest cover change patterns over these two sites, and their correlation with institutional indicators, allows for a process-oriented understanding of forest governance. Therefore, apart from the RS/GIS-based analysis, we also attempt to uncover the socio-economic processes underlying the forest change through interviews and observations.

Analysis of Forest Change over Time Using Satellite and Ground-level Data

The satellite data that we used for this study comprised three multispectral remotely-sensed images from Landsat TM, ETM+ and OLI sensors dated 7 January 1994, 18 January 2003 and 19 February 2014. We chose the data from January and February because they represent the driest period in the landscape and therefore provided maximum contrast. Such contrast enabled us to improve the detection of different vegetation and land use types in the landscape. Details about the potential and capabilities of such multi-spectral Landsat data and their use in mapping vegetation and land use has been extensively reviewed by Roy et al. (2014) and others.[11] All the three images were geometrically corrected using ground control points that we obtained during our extensive field work in the region.

The LULC maps were prepared for the seven classes that we identified from our survey of the landscape. These classes were: 1) Dense forest (old-growth late successional evergreen forests with canopy cover greater than 50 per cent), 2) Degraded forest (secondary forests at various stages of succession with canopy cover between 20 and 50 per cent), 3) Plantation and Agriculture (PAG) (swidden farms, betel nut plantations and orchards), 4) Open area (active and abandoned open cast limestone mines along with granite quarries), 5) Water, 6) Settlement (built-up area), 7) Industry (cement-manufacturing units). Details about accuracy estimate of the classification and the techniques used to determine post-classification change detection using the matrix tool of ERDAS IMAGINE has been provided in Goswami et al. (2016).

The boundary of the RF was sourced from the FD. The *elaka* boundary was demarcated in consultation with the respective village and the *elaka* council members. The area under each LULC class was extracted for the *elaka* and the RF for comparison. The relative proportion of each class with respect to the total area of the *elaka* and the RF were calculated for all the three years.

Patterns of Forest Change

The RS and GIS-based analysis showed an increase in the degraded forest area, open area, plantation agriculture and industry, and a decrease in dense forest area from 1994–2014. This change pattern was similar across both the RF and the CF across both the time-periods—1994–2003 and 2003–14—for which the analysis was carried out. However, it was the magnitude of these changes that differed vastly between the CF and the RF, particularly during the second decade (from 2003–14). The magnitude of change was much higher in the CF, leading to partial transformation of the forest-agro-plantation-dominated land use to extractive mining and industrial land use.

Till 1994, the landscape was dominated by dense forests across both the RF (87 per cent) and the CF (72 per cent), which remained almost unchanged till 2003. However, by 2014, while the RF still maintained 75 per cent of its landscape as dense forests, only 27 per cent of the CF had dense forest cover (see Figure 8.2).

Figure 8.2: Graph Indicating Area of the Different LULC Classes

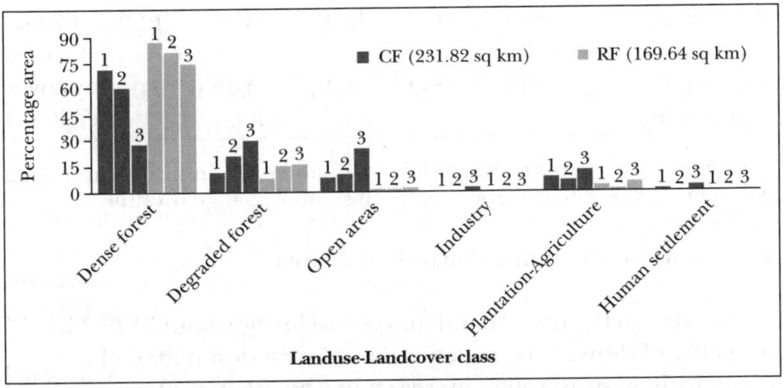

Note: Area in relative percentage across CF and RF for 1994, 2003 and 2014, indicated as 1, 2 and 3, respectively, above the bars. Figures in parenthesis beside 'CF' and 'RF' indicate their total extent in sq. km.

Source: Based on the authors' analysis of RS data.

Table 8.1 shows that bulk of the forest loss in the CF took place between 2003 and 2014, with considerable gains in degraded forest area, open area (mining), industry (chiefly cement-based industry) and plantation agriculture. All of these categories occupied negligible proportions of the CF till 2003. Alongside, the 'human settlement' category added 6.25 sq. km between 2003 and 2014.

The analysis showed that during this time, the CF transformed rapidly from a forest-based and subsistence agriculture-dominated landscape to an extensive production-oriented landscape, which included mining, industries and betel nut plantations. Further, a large area of degraded forest was also added during this period.

Table 8.1: Relative Percentage of LULC Categories Across CF and RF for 1994, 2003 and 2014

LULC Categories	*CF (231.82)*			*RF (169.64)*		
	1994	*2003*	*2014*	*1994*	*2003*	*2014*
Dense Forest	71.70	60.92	27.30	87.15	81.59	74.82
Degraded Forest	10.36	20.82	29.91	8.03	15.73	16.70
Open Areas	8.13	9.79	24.79	0.70	0.80	2.14
Industry	0.00	0.27	2.39	0.00	0.04	0.01
Plantation Agriculture	8.07	6.77	12.15	3.01	1.30	5.13
Human Settlement	0.11	0.41	3.11	0.00	0.03	0.27

Note: Figures in parenthesis beside 'CF' and 'RF' indicate their total extent in sq. km. The categories with substantial change over the years are highlighted.

Source: Based on the authors' analysis of RS data.

The RF, on the other hand, in spite of losing about 21 sq. km (14 per cent) of dense forests, retained its forest-dominated character, albeit with a substantial increase in the plantation agriculture acreage between 2003 and 2014. Dense forest was mainly converted to degraded forest (which doubled in area) and to negligible areas of open areas and settlements.

The main driver behind the four-fold increase in the plantation agriculture category within the RF is the expansion of existing plantations and the setting up of new ones, particularly *pan-jhum* plantations (betel leaf plantations) within the RF. *Pan-jhum's* acreage has increased due to a rapid increase in both the demand for and the market price of the leaf variety cultivated in this region. Due to its size, quality and taste, the leaf is marketed all over the state as well as in the neighbouring states of Assam, West Bengal, Tripura and Mizoram, and Bangladesh, all of where chewing betel nut is inherently rooted in their cultures.

Drivers of Forest Change in State and Community Forests

The change matrix analysis (Appendices 8.1 and 8.2) showed that a large area of dense and degraded forest has been converted to non-forest use in the CF between 2003 and 2014. Out of the total 77.94 sq. km for dense forest lost in the CF during this period, 34.54 per cent (26.92 sq. km) were converted to open (mining) class, 17.33 per cent (13.55 sq. km) to plantation and agriculture, and 4.7 per cent (33.68 sq. km) to settlements. For the same period, 28.10 per cent, 16.8 per cent, and 2.6 per cent of the degraded forest class were converted respectively to open, plantation agriculture and human settlement in the CF.

In comparison, only 1.78 sq. km (equal to 1.3 per cent) of dense forests were converted to open class in the RF between 2003 and 2014. Here bulk of the loss, 6.1 sq. km out of the total 11.73 sq. km or 4.4 per cent, was accounted for by the conversion of primary forest to plantation agriculture. There was negligible conversion of the dense forest category to human settlements or to industries.

Large-scale Drivers of Forest Loss—The Policy Push

Clearly the highest forest loss in the community-controlled areas was driven by limestone mining and industries. The southern slopes of the Meghalaya plateau, where the study area is situated, contain the largest reserves of high-grade limestone, ideal for cement manufacturing.[12] However, the trigger for its exploitation at a massive scale was the need to maintain high economic growth rates, which the liberalisation of the economy unleashed in the early 1990s. To maintain such high growth rates, it was imperative that India explored its developmental frontiers, such as the north-eastern region, which was highly endowed with natural and mineral resources such as forest, water (hydropower), coal, oil, limestone and uranium. However, the inherent rights of the local communities over the majority of the land and its resources (which is over 70 per cent of the total land area) coupled with protective laws such as Articles 244 and 275 of the Sixth Schedule and the Meghalaya Transfer of Land (Regulation) Act, 1971[13] hampered the 'free climate' that large industries required to thrive. Moreover, the large extent of forests in the region necessitated forest clearance. Yet each of these hurdles to rapid economic growth was tackled with new policies or a modification/ amendment of the existing policy.

On 24 December 1997, writing to all the northeastern states and ministries about the formulation of the North-East Industrial Investment Promotion Policy (NEIIPP), Ashok Kumar, Joint Secretary in the Ministry of Industry, said:

> In view of the continuing backwardness of North East region the need for a new and synergetic incentive package was widely felt to stimulate development of industries (Hon'ble Prime Minister made a statement at Guwahati on October 27, 1996 that new initiatives would be announced for the industrial development of the North Eastern Region) [14]

A thorough review of all the policies implemented in the last 30 years revealed NEIIPP to be the key driver behind the rapid surge in the cement-based industries and limestone mining. This policy eased regulations, and created massive subsidies on loans, transportation and investments. Within the next decade, seven large cement factories were set up in the CF alone. Such a policy was in tune with the prioritisation of economic growth over everything else by the post-neoliberal state of India which viewed the northeast as a vast but 'unruly' and 'untapped' reservoir of valuable natural and mineral resource.[15] Such rapid growth of industry was facilitated by the dilution of existing laws and regulations related to land rights. For example, almost all the hill states of the northeast which come under the jurisdiction of the Sixth Schedule have well-defined land protection acts and regulations, which prevent non-tribals from owning land in these areas. In Meghalaya, for example, the Meghalaya Land Transfer Act, 1971 was the key legislation which prevented land hoarding by non-tribal people in the tribal-dominated areas.

However, to facilitate extraction of resources, setting up of industries and transferring of land to non-tribal entities, such acts were diluted. In this case, the Meghalaya Land Transfer Act, 1971 was revised through an executive order of 2001, which said that land may be transferred to a non-tribal entity if it is in 'tribal interest' (Pyrtuh 2012). It is therefore hardly surprising to learn that all the cement factories and other limestone-allied industries that have appeared in the landscape in the last 10 years are owned by large industrial conglomerates, one of which was a French multinational giant which acquired thousands of acres of forest land.[16] Further, in order to bypass the requirement of forest clearance, the definition of forest was revised in Meghalaya on 28 December 2012 by amending the Forest Regulation Act, 1973.[17] According to the new definition, only a 'compact and continuous tract of minimum four hectares' of vegetation will now be considered as 'forest' in Meghalaya. To put this definition into context, there are many sacred forests in Meghalaya which are less than 4 hectares in size (Tiwari et al. 1998), all of which cease to be 'forest' following the definition. Moreover, owing to mining and setting up of industries, many large forest areas are now converted to small fragments. All such areas can now be easily obtained for mining and industrial operations without any forest clearances.

Such modifications in policy are not merely restricted to Meghalaya. Elsewhere in the northeast, too, new policies are being introduced to facilitate large-scale oil palm plantation in Mizoram, Joint Forest Management Committee (JFMC) schemes in Manipur, oil exploration and drilling in Nagaland and setting up of large dams in Arunachal Pradesh (Shrivastava 2015).

Role of Plantation-Agricultural Expansion

Plantation-agriculture land use was the second largest driver of forest loss in the landscape of the CF. As the change matrix data shows, plantation agriculture contributed to the loss of 7.89 sq. km (11.4 per cent) of forest in the RF and 21.63 sq. km (26.4 per cent) in the CF between 2003 and 2014 (see Appendices 8.1 and 8.2). An elaboration of this category is warranted here. *Jhum* (shifting) cultivation was the traditional form of agriculture in the southern and south-eastern hilly landscape of Meghalaya, which had an average fallow cycle of 7–11 years in the Jaintia Hills (Goswami 2015). Abundant forest area with low population density helped maintain the forest-fallow and *jhum* dynamic in the landscape, the scale of which was mostly subsistence in nature with the area cleared being not more than 0.1–1 hectare in size. As Metzger (2002) argued, slash-and-burn landscape remains sustainable if the conversion of forest to agriculture may be compensated by forest regeneration in a way that primary and secondary forest reserves may be maintained over time.

As our data shows, during 1994–2003, the plantation-agriculture growth rates were almost zero, while the rates of 'dense forest' loss and increase in 'degraded forest' were identical, indicating the maintenance of an equilibrium in terms of the overall forest acreage. This indicates that the predominant *jhum* cultivation, carried out with longer fallow periods, were somewhat sustainable in nature, which helped to maintain the forest landscape in a 'steady-state condition' (Turner et al. 1993). However, during the last decade, the rapid shift in the landscape from *jhum* to permanent cash crops, chiefly areca nut (*Areca catechu*), seem to have driven the high growth rates in plantation-agriculture land use category. Areca nut, which has high demand in the local and international markets, are maintained as monoculture stands for 30–50 years, following which there is a change in the soil characteristics (due to increased run-off since undergrowth is cleared). In such areas, forest regeneration is difficult and slow according to the local planters, and long-term observations also show this category of crop area to be the least diverse in terms of bird and butterfly diversity. Such fine-scale changes and a shifting preference for cash crop rather than subsistence-scale agriculture might be the insidious fine scale local process, which is resulting in the upsetting of the forest-fallow-*jhum* dynamic.

Similar shifting patterns from traditional food to cash crops have also been reported from other regions of the northeast and the rest of the country. Behera et al.'s study (2016) on the plateau region of Meghalaya showed how a five-fold increase in the market price of broom-grass (*Thysanolaena maxima*) has pushed people into abandoning traditional food crops which provided food-security in the past. Such shifts were particularly large-scale during the last decade and were facilitated by increasing demand, better linkages to the market, improvement in supply-chain networks and surface-transport infrastructure. Similarly, manifold rise in the market price of the Malnad variety of betel nut led to the conversion of *kan* sacred forests to betel nut plantations (Gokhale 2004). *Kan* are evergreen forest patches in the Western Ghats in Karnataka, which are protected by the people due to religious values.

Another less evident driver of such increasing acreage of plantation (permanent and semi-permanent plantation) is embedded within the process of conversion of communal land to private property. As per the customary rules of the *Pnars* (the indigenous and dominant tribal community of Jaintia Hills), a person who makes permanent changes on allotted common land is entitled to claim private ownership over it. For example, planting multi-seasonal trees like pine, betel nut or orange, growing broom-grass, creating fish pond, or erecting a building on the allotted land can be claimed for private ownership, while a *jhum* fallow abandoned after cultivation and harvest returns to the communal land pool (Cantlie 1934).

Following the rapid proliferation of mining, industrialisation and the ancillary development such as improved road networks, emergence of new settlements, and markets and population growth, there has been a sharp increase in the value of and demand for land. Such changes have led to an increase in privatisation of land by acquiring titles from the *Doloi* and District Council, usually through the establishment of multi-seasonal plantations and permanent structures (as stated earlier) on land that was earlier communally owned. Such changes in the land tenure regimes may also have played an important role in the rapid shift from small-scale *jhum* to large-scale plantations, thereby resulting in irreversible land use changes.

Forest Loss–Governance Linkages

Previous meta-analyses of conservation effectiveness report policy, market pressure, connectivity, human population density, developmental and economic pressures, and institutional arrangements as the key variables impacting the outcomes (Ellis and Porter-Bolland 2008). In our case, such variables were similar across the CF and the RF, given that both were situated within a similar geopolitical landscape. However, institutional arrangements (forest laws, regulations, forest management bodies and their interactions) varied across the CF and the RF; they are considered here in order to explain some of the differences in the forest conversion trajectories.

In the RF, in tune with the existing forest laws and regulations and local management priorities, the FD has not permitted mining and industries within its boundaries but has chosen to allow small-scale agricultural activities for the forest-dependent communities. Moreover, most ground staff is part of the same community which are cultivating within the RF area. Therefore, rather than upsetting social relations, they allow such transgressions within the RF as long as they have older antecedents. More recently, with the establishment of a new wildlife sanctuary in 2014 out of the erstwhile 168 sq. km RF, the forest boundaries have come to be strictly enforced and monitored and new cultivation is being discouraged. The frequency of management and protection-related activities (such as patrolling, boundary marking, etc.) has increased considerably partly due to improved access to funds. Yet, as the change matrix shows, an increasing area of agriculture and monocultural plantation is appearing within the RF and the future trajectory of this change is uncertain.

In the CF, however, our observations suggest confusion regarding the application of existing forest laws and regulations within the CF and village forests. The retaining of the JHADC, which maintains its own independent forest rules and regulations, even after attaining full statehood, adds another institutional layer to the prevailing confusion. Additionally, conflict has been reported between the traditional institutions, the ADCs and the FD regarding control and management of forests and its resources. Such confusion due to the presence of multiple institutions might have facilitated the unregulated growth of mining and industries. According to the customary laws of the Jaintia Hills, the decision to allow mining and industries in the community land required the permission of the *doloi* and the approval of the respective village council(s) within which the mines/industries are located, along with a formal approval by the JHADC. In this context, it is important to note that most of the mines and industries are set up through merely obtaining no-objection certificates (NOCs) from the *doloi* and *waheh chnongs*. This implies that decisions that were to be taken by a three-tier structure are being subverted by this simplified process of approval. A growth-friendly political dispensation further exacerbates the situation. Further, the small size of most of such forest diversions precludes the need for reference to the central Ministry of Environment, Forest and Climate Change (MoEFCC), and thus escapes the formal Environmental Impact Assessment (EIA) process.

The cash-strapped local communities appear to see industries and mining as necessary means of access to the development of basic infrastructure such as roads and availability of jobs, and therefore, seem to overlook changes that may affect their forest-based livelihoods. For example, there is a loss of water sources for several villages in the vicinity of the cement factories, where a lot of forests have been cleared for mining (pers. obs.). However, our discussions indicate that with increased access to cash incomes, people are now buying water from tankers. Improvement in secondary and tertiary road networks have also led to easier access to markets, which might have facilitated an expansion of cash crop acreage.

Conclusion

Through this chapter, we showed how the development and economic growth trajectory of the northeast came to a head following the liberalisation policies unleashed by the Indian government in 1991. The region that was once the hub of economic activity during the colonial regime, owing to the oil, coal, timber and tea industry, was cordoned off post-1947 from the mainstream path of growth and development following Nehru's policy on the tribal areas of the northeast. However, today, after 25 years of economic reforms, the region is once again at the centre of India's overly development-oriented policies, which seek to realise its immense potential in hydropower,[18] crude oil, coal, commercial plantations, uranium,[19] and military and civil infrastructure.[20] Due to its strategic location and the Indian government's keenness on attracting foreign investment, the northeast has also become an important site for countries seeking to control China's dominance in the Southeast Asian region.[21, 22]

However, as our case study of the Jaintia Hills demonstrated, such interactions of global trade demands and regional economic policies incorporated one of the remotest corners of Meghalaya within the newly adopted industrial and mining economy. Such incorporations have come with relatively large impacts on the natural forests of the Jaintia Hills and have been aided by the weakening of village-level institutions. The presence of the state-managed RF seemed better able to withstand the pressures of mining and industrialisation, although these forests are still vulnerable to agricultural expansion. Clearly the pace of land use change is far lower in state-governed forests than in community-controlled forests, although the direction of change is similar. In the current situation, the future of forest conservation in Meghalaya would rely on how both the state and the community institutions deal and adapt with the regional pressures of economic growth and development, heavily dependent on natural resources extraction, which have engulfed the rest of the country.

Notes

1. Press Trust of India (PTI), 2017, 'PM Modi Names India's Longest Bridge After Assam Singer Bhupen Hazarika', *NDTV*, 26 May 2017, available at http://www.ndtv.com/india-news/pm-modi-names-indias-longest-bridge-after-assam-singer-bhupen-hazarika-1704256 (accessed on 12 February 2018).
2. The term 'northeast' was believed to have been coined in 1971, with the formation of the North-East Council (NEC), a nodal agency dedicated solely to the affairs and development of northeast India, and was recently upgraded to a separate ministry named the Ministry of Development of North Eastern Region. 'Its vision is to accelerate the pace of socio-economic development of the Region so that it may enjoy growth parity with the rest of the country', claims its website, available at http://mdoner.gov.in/ (accessed on 12 February 2018).
3. The term 'mainland' is used to denote the rest of the country with respect to the north-eastern region comprising the eight states of Assam, Arunachal Pradesh, Manipur, Nagaland, Mizoram, Tripura, Meghalaya and Sikkim. This dichotomy of 'northeast' and 'mainland' is evoked in both scholarly as well as non-scholarly writings to emphasise the spatial and socio-economic marginalisation of the northeast, largely attributed to its unique geo-cultural position and characteristic, existing as a separate entity from what is perceived as the 'mainstream' Indian polity and culture. For example, see S. Baruah, 2005, 'A New Politics of Race: India and its Northeast', *IIC Quarterly* 32(2 and 3): 165–76 (New Delhi: India International Centre).
4. P. Phanjoubam, 2016, *The Northeast Question: Conflicts and Frontiers*, p. 185 (New Delhi: Routledge).

5. The Sino-Indian war was a tip off point wherein China invaded several hundred miles in Arunachal Pradesh but withdrew almost unilaterally. A recent escalation of tensions regarding China's dispute over the MacMohan Line and Arunachal Pradesh, which it claims to be the territory of Southern Tibet, was when the Chinese Army started building a road in the Doklam plateau in the tri-unction of Bhutan, India and Tibet, and the Indian Army prevented it by engaging the Chinese Army in an armless scuffle.
6. In the 1970s and 1980s, the main focus of reorganisation was India's northeast, which was already carried out in the southern region, where intense regional movements started immediately after Independence. Reorganisation was used to contain tribal insurgency and separatist movements through the North-eastern States Reorganisation Act, 1971, which upgraded the Union Territories (UTs) of Manipur and Tripura, and the sub-state of Meghalaya to full statehood, and Mizoram and Arunachal Pradesh (then Tribal Districts) to Union Territories, which also gained full statehood by 1986 and effectively marked the end of a decade-long violence in several of those regions.
7. For example, various military-led operations against various insurgents in Assam, Tripura and Nagaland. Hazarika (1994) gives an authoritative account of many of these operations along with Bhaumik (2009), which also discuss the implementation of draconian laws such as the Armed Forces (Special Powers) Acts (AFSPA) and the toll it took on the basic human rights of the people.
8. M. K. Datta and I. Das, 2017, 'Economic Performance of the Northeastern Region in the Post-liberalisation Period, in D. K. Mishra, and V. Upadhyay (eds), *Rethinking Economic Development in Northeast India: The Emerging Dynamics*, pp. 50–68 (London: Taylor & Francis).
9. Meghalaya website, available at https://www.lonelyplanet.com/india/northeast-states/meghalaya/introduction, *Lonely Planet* (accessed on 12 February 2018).
10. Department of Mining and Geology, Directorate of Mineral Resources, Government of Meghalaya, available at http://megdmg.gov.in/minerals.html (accessed on 12 February 2018).
11. R. C. Estoque and Y. Murayama, 2015, 'Classification and Change Detection of Built-up Lands from Landsat-7 ETM+ and Landsat-8 OLI/TIRS Imageries: A Comparative Assessment of Various Spectral Indices', *Ecological Indicators* 56: 205–17; Y. Ke, J. Im, J. Lee, et al., 2015, 'Characteristics of Landsat 8 OLI-derived

NDVI by Comparison with Multiple Satellite Sensors and In-situ Observations', *Remote Sensing of Environment* 164: 298–313; X. Li, P. Gong, and L. Liang, 'A 30-year (1984–2013) Record of Annual Urban Dynamics of Beijing City Derived from Landsat Data', *Remote Sensing of Environment* 166: 78–90; A. Mei, C. Manzo, G. Fontinovo, et al., 2016, 'Assessment of Land Cover Changes in Lampedusa Island (Italy) using Landsat TM and OLI Data', *Journal of African Earth Sciences* 122: 15–24; X. She, L. Zhang, Y. Cen, et al., 2015, 'Comparison of the Continuity of Vegetation Indices Derived from Landsat 8 OLI and Landsat 7 ETM+ Data among Different Vegetation Types', *Remote Sensing* 7(10): 13485–506.
12. Department of Mining and Geology, available at http://megdmg.gov.in/minerals.html.
13. Meghalaya Act I of 1972, 'The Meghalaya Transfer of Land (Regulation) Act, 1971', available at http://megrevenuedm.gov.in/acts/land-transfer-act-1971.pdf (accessed on 13 February 2018).
14. Anonymous, 2007, 'North Eastern Industrial and Investment Promotion Policy NEIIPP, 2007', Ministry Industry (ed.) (New Delhi: Government of India).
15. K. S. Shrivastava, 2015, 'Battle over Oil, Coal and Forests', *Down to Earth,* 18 September 2015, available at http://www.downtoearth.org.in/coverage/battle-over-oil-coal-forests-46903 (accessed on 14 February 2018).
16. Lafarge Umiam Mining Pvt. Ltd., available at http://www.lumpl.com/about-us/company-profile/ (accessed on 14 February 2018).
17. S. Juneja, 2013, 'Meghalaya Redefines Forest', *Down to Earth,* 31 January 2013 (New Delhi: Centre for Science and Education), Available at http://www.downtoearth.org.in/news/meghalaya-redefines-forest-40048 (accessed on 14 February 2018).
18. Shripad Dharmadhikary, 2008, 'Massive Dam Plans for Arunachal', *India Together,* 17 February 2008, available at http://indiatogether.org/arunachal-environment (accessed on 15 February 2018).
19. B. G. Karlsson, 2011, *Unruly Hills: Nature and Nation in India's Northeast,* p. 336 (New Delhi: Social Science Press and Orient BlackSwan).
20. M. D. Kukreja, 2016, 'Transportation Infrastructure in the North East', *Indian Defence Review* 3: 1–17.
21. K. Bhattacharjee, 2017, 'Japan Keen on Friendship with Northeast', *The Hindu,* 25 September 2017.
22. Elizabeth Roche, 2017, 'India and Japan will Jointly Develop Infrastructure, Tourism, Food Processing and Collaborate in Skill Development in the North-East States', *Livemint,* 4

August 2017, available at http://www.livemint.com/Politics/QZJxtszZgT41umRKoGnBnJ/Govt-plans-Rs45000-crore-investment-in-northeast-states.html (accessed on 15 February 2018).

References

Allen, B. C. 1905. 'Gazetteer of the Khasi and Jaintia Hills, Garo Hills, Lushai Hills'. Shillong: Government of British India.

Anonymous. 1995. 'Management Plan of Narpuh Block 1, 2 and Saipung Reserve Forest'. Jowai: Department of Forest and Environment, Meghalaya.

———. 2007. 'North Eastern Industrial and Investment Promotion Policy NEIIPP, 2007'. New Delhi: Ministry of Industry, Government of India.

———. 2008. 'North Eastern Region Vision 2020'. New Delhi: Ministry of Development of the North Eastern Region and North Eastern Council.

Baruah, S. 2003. 'Nationalizing Space: Cosmetic Federalism and the Politics of Development in Northeast India', *Development and Change* 34(5): 915–39.

———. 2005. 'A New Politics of Race: India and its Northeast', *IIC Quarterly* 32(2 and 3): 165–76. New Delhi: India International Centre.

Behera, R. N., D. K. Nayak, P. Andersen, et al. 2016. 'From Jhum to Broom: Agricultural Land-use Change and Food Security Implications on the Meghalaya Plateau, India', *Ambio: A Journal of the Human Environment* 45(1): 63–77.

Bhattacharjee, K. 2017. 'Japan Keen on Friendship with Northeast', *The Hindu*, 25 September 2017.

Cantlie, K. 1934. *Notes on Khasi Law*, p. 201. Shillong: Chapala Publishing House.

Champion, H. G., and S. K. Seth. 1968. *A Revised Survey of Forest Types of India*, p. 404. New Delhi: Government of India.

Choudhury, A. 2013. *The Mammals of North East India*, p. 431. Guwahati: Gibbon Books.

———. 2014. *A Pocket Guide to the Birds of Meghalaya*, p. 160. Guwahati: Gibbon Books and The Rhino Foundation for Nature in Northeast India.

Datta, M. K., and I. Das. 2017. 'Economic Performance of the Northeastern Region in the Post-liberalisation Period'. In D. K. Mishra and V. Upadhyay (eds), *Rethinking Economic Development in Northeast India: The Emerging Dynamics*, pp. 50–68. London: Taylor & Francis.

Department of Mining and Geology, Directorate of Mineral Resources, Government of Meghalaya. Available at http://megdmg.gov.in/minerals.html (accessed 12 February 2018).

Dharmadhikary, Shripad. 2008. 'Massive Dam Plans for Arunachal', *India*

Together, 17 February 2008. Available at http://indiatogether.org/arunachal-environment (accessed 15 February 2018).

Duarah, C. K. 2017. 'Anti-dam Protests Continue in Arunachal Pradesh', *India Climate Dialogue*, 11 January 2017.

Ellis, E. A., and L. Porter-Bolland. 2008. 'Is Community-based Forest Management More Effective than Protected Areas? A Comparison of Land Use/Land Cover Change in Two Neighbouring Study Areas of the Central Yucatan Peninsula, Mexico', *Forest Ecology and Management* 256(11): 1971–83.

Estoque, R. C., and Y. Murayama. 2015. 'Classification and Change Detection of Built-up Lands from Landsat-7 ETM+ and Landsat-8 OLI/TIRS Imageries: A Comparative Assessment of Various Spectral Indices', *Ecological Indicators* 56: 205–17.

Forest Survey of India (FSI) 2013. 'India State of the Forest Report'. Dehradun: FSI.

Gassah, L. S. 1998. *Traditional Institutions of Meghalaya: A Case Study of Doloi and His Administration.* New Delhi: Daya Books.

Gokhale, Y. 2004. 'Reviving Traditional Forest Management in Western Ghats: Study in Karnataka', *Economic and Political Weekly* 39(31): 3556–59.

Goswami, R. 2015. *Forest Cover, Hunting and Animal Abundances across State and Community Forests of Meghalaya, India*, p. 141. Bangalore: ATREE, Manipal University.

Goswami, R., A. Jesudasan, and T. Ganesh. 2012. 'Can Community Rights, Wildlife Conservation and Mining Coexist? The Meghalaya Experience', *The Hindu*, Survey of the Environment 2012.

Goswami, R., and A. Jesudasan. 2012. 'Mined to Death: An Elegy for the Rivers of Meghalaya, *Down To Earth.* New Delhi: Centre for Science and Environment.

Goswami, R., M. J. Gogoi, and S. N. Karimbumkara. 2015. 'Sighting of Branded Yeoman *Algia fasciata fasciata* Felder & Felder, 1860 (Lepidoptera: Nymphalidae: Heliconiinae) in Jaintia and Cachar Hills, Northeastern India', *Journal of Threatened Taxa* 7(11): 7833–35.

Goswami, R., M. Muneeswaran, T. Somendro, et al. 2016. 'Conservation Effectiveness across State and Community Forest: The Case of Jaintia Hills, Meghalaya, India', *Current Science* 111(2): 380–87.

Hazarika, S. 1994. *Strangers of the Mist: Tales of War and Peace from India's Northeast.* New Delhi: Viking, Penguin Books India.

Jensen, J. R. 2000. *Remote Sensing of the Environment: An Earth Resource Perspective*, p. 544. New Jersey: Prentice Hall.

Juneja, S. 2013. 'Meghalaya Redefines Forest', *Down to Earth*, 31 January 2013. New Delhi: Centre for Science and Education. Available at http://www.downtoearth.org.in/news/meghalaya-redefines-forest-40048 (accessed 14 February 2018).

Karlsson, B. G. 2011. *Unruly Hills: Nature and Nation in India's Northeast*, p. 336. New Delhi: Social Science Press and Orient BlackSwan.

Kashyap, S. G. 2016. 'Behind the Firing and Deaths in Tawang: Anti-dam Protests and Aabbot-lama Tussle', *The Indian Express*, 4 May 2016.

Ke, Y., J. Im, J. Lee, et al. 2015. 'Characteristics of Landsat 8 OLI-derived NDVI by Comparison with Multiple Satellite Sensors and In-situ Observations', *Remote Sensing of Environment* 164: 298–313.

Kukreja, M. D. 2016. 'Transportation Infrastructure in the North East', *Indian Defence Review* 31: 1–17.

Lafarge Umiam Mining Pvt. Ltd. Available at http://www.lumpl.com/about-us/company-profile/ (accessed 14 February 2018).

Lambin, E. F., and P. Meyfroidt. 2010. 'Land Use Transitions: Socio-ecological Feedback Versus Socio-economic Change', *Land Use Policy* 27(2): 108–18.

Li, X., P. Gong, and L. Liang. 2015. 'A 30-year (1984–2013) Record of Annual Urban Dynamics of Beijing City Derived from Landsat Data', *Remote Sensing of Environment* 166: 78–90.

Mei, A., C. Manzo, G. Fontinovo, et al. 2015. 'Assessment of Land Cover Changes in Lampedusa Island (Italy) using Landsat TM and OLI Data', *Journal of African Earth Sciences* 122: 15–24.

Meghalaya Act I of 1972. 'The Meghalaya Transfer of Land (Regulation) Act, 1971'. Available at http://megrevenuedm.gov.in/acts/land-transfer-act-1971.pdf (accessed 13 February 2018).

Meghalaya website. Available at https://www.lonelyplanet.com/india/northeast-states/meghalaya/introduction, *Lonely Planet* (accessed 12 February 2018).

Metzger, J. P. 2002. 'Landscape Dynamics and Equilibrium in Areas of Slash-and-Burn Agriculture with Short and Long Fallow Period (Bragantina Region, NE Brazilian Amazon), *Landscape Ecology* 17(5): 419–31.

Ministry of Development of North Eastern Region. Available at http://mdoner.gov.in/ (accessed 12 February 2018).

Munyati, C., and T. A. Kabanda. 2009. 'Using Multitemporal Landsat TM Imagery to Establish Land Use Pressure Induced Trends in Forest and Woodland Cover in Sections of the Soutpansberg Mountains of Venda Region, Limpopo Province, South Africa', *Regional Environmental Change* 9(1): 41–56.

Papola, T. C. S. 2017. 'Globalisation and the Himalayan States: Mitigating or Accentuating Marginalisation?' In D. K. Mishra and V. Upadhyay (eds), *Rethinking Economic Development in Northeast India: The Emerging Dynamics*, pp. 32–49. London: Taylor & Francis.

Parrish, J. D., D. P. Braun, and R. S. Unnasch. 2003. 'Are We Conserving What We Say We Are? Measuring Ecological Integrity within Protected Areas', *BioScience* 53(9): 851–60.

Phanjoubam, P. 2016. *The Northeast Question: Conflicts and Frontiers*, p. 2019. New Delhi: Routledge.

Press Trust of India (PTI). 2017. 'PM Modi Names India's Longest Bridge After Assam Singer Bhupen Hazarika', *NDTV*, 26 May 2017. Available at http://www.ndtv.com/india-news/pm-modi-names-indias-longest-bridge-after-assam-singer-bhupen-hazarika-1704256 (accessed 12 February 2018).

Pyrtuh, P. 2012. Desolation of the Meghalaya Land Transfer Act, *Shillong Times*, 18 October 2012.

Roche, Elizabeth. 2017. 'India and Japan will Jointly Develop Infrastructure, Tourism, Food Processing and Collaborate in Skill Development in the North-east States', *Livemint*, 4 August 2017. Available at http://www.livemint.com/Politics/QZJxtszZgT41umRKoGnBnJ/Govt-plans-Rs45000-crore-investment-in-northeast-states.html (accessed 15 February 2018).

Roy, D. P., M. A. Wulder, T. R. Loveland, et al. 2014. 'Landsat-8: Science and Product Vision for Terrestrial Global Change Research', *Remote Sensing of Environment* 145: 154–72.

Saikia, A. 2011. *Forests and Ecological History of Assam: 1826–2000*, p. 388. New Delhi: Oxford University Press.

Sharma, I., and K. S. Shrivastava. 2011. 'Lafarge's Fait Accompli', *Down to Earth*, 31 July 2011.

She, X, L. Zhang, Y. Cen, et al. 2015. 'Comparison of the Continuity of Vegetation Indices Derived from Landsat 8 OLI and Landsat 7 ETM+ Data among Different Vegetation Types', *Remote Sensing* 7(10): 13485–506.

Shrivastava, K. S. 2015. 'Battle over Oil, Coal and Forests', *Down to Earth*, 18 September 2015. Available at http://www.downtoearth.org.in/coverage/battle-over-oil-coal-forests-46903 (accessed 14 February 2018).

Terborgh, J., and C. P. van Schaik. 2002. 'Why the World Needs Parks'. In J. Terborgh, C. van Schaik, L. Davenport, and M. Rao (eds), *Making Parks Work: Strategies for Preserving Tropical Nature*, pp. 3–14. USA: Island Press.

Tiwari, B. K., S. K. Barik, and R. S. Tripathi. 1998. 'Sacred Groves of Meghalaya'. In S. Ramakrishnan, K. G. Saxena, and U. M. Chandrasekhar (eds), *Conserving the Sacred for Biodiversity Management*. New Delhi: Oxford University Press and IBH Publishing.

Turner, M. G., W. H. Romme, R. H. Gardner, et al. 1993. 'A Revised Concept of Landscape Equilibrium: Disturbance and Stability on Scaled Landscapes', *Landscape Ecology* 8(3): 213–27.

Velho, N, U. Srinivasan, P. Singh, et al. 2016. 'Large Mammal Use of Protected and Community-managed Lands in a Biodiversity Hotspot', *Animal Conservation* 19(2): 199–208.

Appendix 8.1

LULC Transition Matrix for CF (Narpuh *Elaka*) from 2003–14

LULC Class	*Waterbody*	*Industry*	*Settlements*	*Open*	*Plantation and Agriculture*	*Degraded Forest*	*Dense Forest*	*Total 2003*
Water Body	0.09	0.25	0.10	0.93	0.00	0.10	0.03	1.51
Industry	0.00	0.23	0.17	0.65	0.11	0.23	0.09	1.47
Settlement	0.01	0.08	0.22	0.39	0.02	0.17	0.07	0.96
Open Area	0.11	1.52	1.37	11.84	1.47	5.00	1.39	22.69
Plantation and Agriculture	0.01	0.26	0.39	3.19	4.95	5.78	1.12	15.70
Degraded Forest	0.22	1.32	1.27	13.55	8.12	15.07	8.71	48.26
Dense Forest	0.36	1.89	3.68	26.92	13.51	42.99	51.87	141.22
Total 2014	0.80	5.54	7.21	57.47	28.17	69.34	63.28	231.82

Note: Each row in the matrix indicates the area of the original LU (left most and bold) to other LUs. All figures are absolute area and are expressed as sq. km.

Source: Based on the authors' analysis of RS data.

APPENDIX 8.2

LULC Transition Matrix for RF (Narpuh RF) across 2003–14

LULC Class	*Waterbody*	*Industry*	*Settlements*	*Open*	*Plantation and Agriculture*	*Degraded Forest*	*Dense Forest*	*Total 2003*
Waterbody	0.23	0.00	0.03	0.15	0.00	0.06	0.02	0.49
Industry	0.03	0.00	0.01	0.02	0.00	0.01	0.01	0.07
Settlement	0.00	0.00	0.00	0.02	0.00	0.03	0.01	0.06
Open area	0.08	0.01	0.05	0.44	0.05	0.55	0.18	1.35
Plantation and Agriculture	0.01	0.00	0.01	0.03	0.77	0.73	0.68	2.22
Degraded Forest	0.60	0.00	0.16	1.19	1.79	8.89	14.16	26.79
Dense Forest	0.64	0.00	0.20	1.78	6.10	18.07	111.88	138.66
Total 2014	1.59	0.01	0.45	3.63	8.70	28.33	126.93	169.64

Note: All figures are absolute area and are expressed as sq. km. Each row in the matrix indicates the area of the original LU (left most and bold) to other LUs.

Source: Based on the authors' analysis of RS data.

9

Hunting Stories and Shady Tales

The Impact of Legislation on Wildlife Conservation and Tree Preservation in the Western Ghats, India*

Archana Bali# and Kartik Shanker

Beyond Protected Areas

Till recently, the modern conservation movement was largely based on two dominant strategies—reservation and restoration (Rosenzweig 2003; Western 2001); reservation prevents further area from being degraded, while restoration returns degraded areas to a state of naturalness.

However, protected areas (PAs) cover less than 12 per cent of the earth's land surface and are inadequate in representing the planet's species diversity (Rodrigues et al. 2004). Tropical forests, which harbour much of the earth's biodiversity, are also severely fragmented (Food and Agriculture Organization [FAO] 2006). Many biologists have also argued that the linear relationship between species richness and area constrains the number of species that can be present in PAs alone (Rosenzweig 2001).

* We are very thankful to the coffee planters of Chikmagalur region and Karnataka Forest Department for participating in this research; the Wildlife Conservation Society-India Programme for funding this study; Dr Ajith Kumar for his support and encouragement and Dr Esha Shah for her valuable inputs in the study design. We are also grateful to K. Sivaramakrishnan, Arun Agrawal, Gary Kofinas and Meera Anna Oommen for their comments on an earlier version of the manuscript.

Archana Bali, wildlife conservationist and biologist, passed away in 2014.

Therefore, conservation strategies that entirely rely on remnant patches (in this case, the PAs) to preserve diversity are unlikely to be successful in the long-term if sufficient attention is not paid to the surrounding matrix. Additionally, given current human population sizes and growth rates, it may not be feasible or fair to substantially increase the area under reservation or restoration, since many biodiversity-rich areas coincide with areas where the world's poorer communities live.

Hence, conventional strategies are clearly inadequate to conserve biodiversity, and long term and large-scale conservation may require looking beyond the boundaries of PAs (Gardener et al. 2009; Janzen 1998; Vandermeer and Perfecto 2007). Human-dominated areas, particularly matrices which include both natural areas and agricultural lands, are set to play an increasingly relevant role in long-term biodiversity conservation. Recent research on the socio-political ramifications of the protectionist approach also stress the importance of adopting alternate, participatory conservation arrangements (Brockington 2001).

The agricultural matrix has specifically received a lot of attention in this regard since 2000 as an opportunity for the protection of biodiversity, and rightly so since it covers 38 per cent of land worldwide (United Nations Environment Programme [UNEP] 2001). While there are numerous programmes being run by conservation organisations worldwide to promote forest-friendly and sustainable agriculture as a means to protect tropical trees and birds, conservationists are divided on whether these approaches work.

A Coffee-table Story from the Western Ghats

The Western Ghats is one of the eight hottest biodiversity hotspots in the world in terms of the number of endemic plants and vertebrates as well as the number of endemic plants and vertebrates per unit area (Myers et al. 2000), but less than 10 per cent of its approximately 1,60,000 sq. km area is set aside for conservation in form of PAs. It has more than 1,000 vertebrate species and nearly 5,000 plant species, including many endangered mammals such as the lion-tailed macaque, the Nilgiri tahr and the grizzled giant squirrel, as well as significant global populations of the Asian elephant and tiger. It is also a well-known hotspot for herpetofauna, with over 80 per cent of frogs and 60 per cent of reptiles being endemic to the region.

Given the biodiversity of the Western Ghats, the area set aside for its conservation is inadequate for several reasons. In fact, many PAs were not originally established with the specific objective of protecting biodiversity. The region has also felt the impact of fragmentation, the major drivers being the conversion of natural forests to plantations of tea, coffee, cardamom and teak, dating back to the 1740s (Elliot 1898; Menon and Bawa 1997).

In the Western Ghats, coffee plantations cover approximately 3,600 sq. km; they are almost entirely shade-grown, and to a large extent, still traditionally managed with a wide variety of spices and fruit crops like vanilla, orange and banana growing alongside (Department of Commerce [DoC] 2009).[1] Throughout the tropics, such traditional agricultural practices which incorporate trees are found to be compatible with the conservation of native biodiversity (Daily et al. 2003; McNeely and Schroth 2006; Moguel and Toledo 1999; Siebert 2002). Coffee plantations in the Western Ghats region provide habitat for some species and increase the overall connectivity of natural habitats considerably (Bali et al. 2007; Das et al. 2006). Given that they provide significant support for biodiversity, these shade coffee plantations provide excellent opportunities for the conservation of wildlife outside PAs (Karanth et al. 2016). Their location in the most productive zones around the middle elevations, strategically enveloping the remnant forests, makes them critical for biodiversity conservation (Das et al. 2006). Thus, these plantations could play an important role in supporting conservation in the region (Anand et al. 2008; Bhagwat et al. 2008; Shahabuddin 1997; Shankar Raman 2006).

In India, conservation is largely viewed and practiced as two dichotomous extremes—conventional top-down state policies for promoting and enforcing conservation, and bottom-up programmes for participation and involvement of local communities. We aimed to evaluate the effectiveness of the top-down approach and identify constraints in achieving conservation objectives in human-dominated landscapes, in terms of the interaction of state policies and local practices (Figure 9.1). Specifically, we asked: how is conservation outcome influenced by local practice, attitudes and state policy? In order to do this, we examined the two laws most relevant to the conservation of plants and animals in the region. Specifically, we looked at hunting in the context of the Wildlife Protection Act, 1972 (Dutta et al. 2004), the most important national legislation which governs the protection

Figure 9.1: Map of Study Area Showing the Location of Coffee Plantations around the Bhadra Wildlife Sanctuary

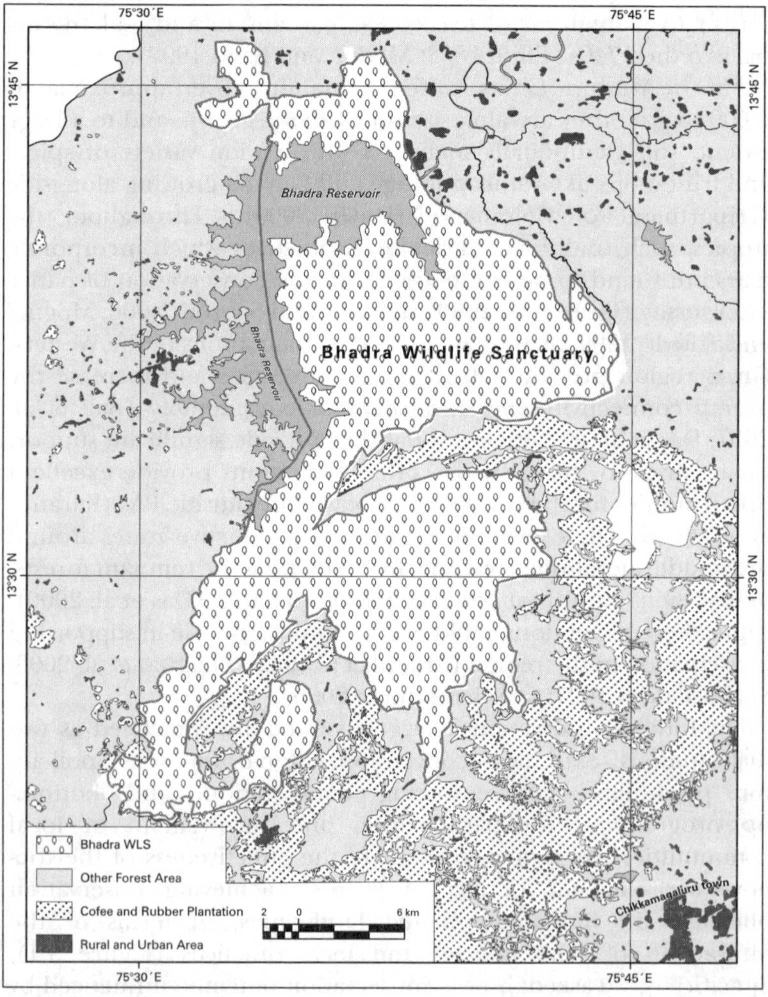

Note: This is a typical example of the juxtaposition of human land use around fragmented forests in the Western Ghats

Source: Study carried out by the authors.

of endangered species and the declaration of PAs (See Appendix 9.1). Second, we examined the Karnataka Preservation of Trees Act (KPTA), 1976 (Karnataka Forest Department [KFD] 2000) and its impact on local planting practice (See Appendix 9.2). We carried out a study (the author Archana Bali carried out this field work as part of her MS dissertation) from December 2005 to May 2006 in Chikmagalur[2] district of Karnataka. We used a qualitative case study approach to evaluate awareness of and compliance with conservation laws by local communities (Appendix 9.1).[3]

History and Politics of the Conservation Laws

During the colonial era, the reservation of forests was primarily aimed at production, natural resource extraction and exercising state control; although there was a limited ethic of conservation behind the reservation of forests and preservation of game; these legislations helped large tracts of forests and wildlife to survive (Rangarajan 2001). The National Forest Policy of India was formulated in 1952 (Ministry of Environment and Forests [MoEF] 2009), whereby state-owned forests were managed by the forest department, but on private lands, the indiscriminate felling and destruction of trees resulted in the loss of timber, and considerable soil erosion in hill areas. To prevent this and to promote the conservation of trees in private areas, the Tree Preservation Order was issued to various states in 1952. This prohibited the felling, lopping or willful destruction of trees. Its main aim was to preserve native trees outside forests. This Order encouraged various states to enact comprehensive tree preservation legislations (Tamil Nadu Hill Areas [Preservation of Trees] Act, 1955; Maharashtra Preservation of Trees Act, 1975; KPTA, 1976; Goa Preservation of Trees Act, 1984; Kerala Preservation of Trees Act, 1986; Delhi Tree Preservation Act, 1994, etc.).

In a nutshell, the KPTA provides for the preservation of trees in the state by regulating the felling of trees and planting of adequate number of trees to restore ecological balance. According to this law, 'preservation of trees' means the maintenance and protection of trees to encourage normal growth and the prevention of damage or cutting or felling. There is a provision for the appointment of a 'tree officer' by the forest department and a 'tree authority' for each urban and rural area, whose duty is to preserve all trees within his/her jurisdiction. The tree officer has the responsibility of carrying out a census of existing trees and specifying the minimum numbers and kinds of trees that each region and land type should have. In specific cases, however, the 'tree authority' can grant permissions for tree felling.

The protection of forests (various categories of PAs, reserved and revenue forests) is the primary responsibility of the State Forest Department. At the field level, the smallest administrative unit is called a 'beat'. A 'Beat-Guard' is responsible for regularly patrolling the beat (often more than 10 sq. km) and preventing any illegal act such as felling of trees, hunting and poaching, grazing, forest fires, etc. In addition to this, the Forest Department is also responsible for the protection of wildlife and native trees in landscapes outside PAs. Forest officers are vested with powers to enquire into offence cases, stop and search vehicles, arrest without warrant, issue search warrants, seize forest produce suspected to be illegally obtained, and confiscate vehicles and other property used to commit forest offences, and so on.

Hunting Stories

Hunting was not uncommon in the region, as revealed by many respondents. A group of plantation labourers provided the following information over an informal discussion:

> Though '*harbeku*' (flying squirrel, *Petaurista philippensis*) is very tasty, it has very little meat and at least 3–4 animals are required for a good meal ... '*Cubbeku*' (common palm civet, *Paradoxurus hermaphroditus*) meat is palatable only before winter; during winter the animal feeds on a diet that lends a foul odor to its meat... '*Kadav*' (sambar, *Cervus unicolor*) meat is very chewy and '*kadu-koorie*' (muntjac, *Muntiacus muntjak*) is preferred to it.... It is easy to catch a '*barka*' (mouse deer, *Moschiola meminna*)—it only has to be chased down a slope and once it's choked, one can simply collect it.

There were two distinct categories of hunters, namely the small subsistence hunters who are estate labourers and villagers, and the sport hunters who are mainly local planters. There were differences in the way these two groups hunted as well as the reasons for which they hunted (Table 9.1). Plantation labourers and villagers hunted for supplementary food and income. The daily wage of a coffee estate labourer ranged from Rs 65–72, less than $2 in 2006, while the price of mutton was roughly around Rs 120–40 kilograms in the local market. Many plantations, especially those close to the Bhadra Wildlife Sanctuary, are also far from the nearest town. In such areas, hunting was also practised opportunistically to procure meat. Labourers often used snares to catch hare and wild boar, smoked porcupine burrows and flying squirrel nests, and caught civet kittens from tree trunks, especially during the annual shade-lopping season; doves and green pigeons were also hunted. Also, certain animals such as monitor lizards, snakes and civets were poached for skin by villagers; however, this was not a common practice. In the villages, a few people owned locally manufactured guns and used them to hunt in small forest fragments. Gun-sharing was a standard practice.

Table 9.1: Differences in Hunting Practices between Planters and Plantation Labourers

	Labourers	*Planters*
Reason for Hunting	For food and supplementary income.	For sport and recreation.
Organisation	Mostly opportunistic hunting. Group size was usually very small and comprises 1–3 individuals.	Hunts were always organised. Group size ranged from 2–6 or more.
Target Species	Usually small animals—civets, small cats, hare and birds.	Mostly large animals—wild boar, sambar, muntjac.
Method of Hunting	Snaring, smoking the burrows, trapping.	Using guns and flashlights.
Area of Activity	Very localised, usually within the plantation and neighbouring areas.	Active within a larger region. Used four-wheeled vehicles such as jeeps to get to remote areas.

Source: Field study by the authors.

On the other hand, the planters hunted mainly for sport. These hunts were generally organised for a group of people, often 4–6 or more. They used guns, four-wheel drive vehicles and spotlights. Hunts were usually organised as 'nights-out' for men on occasions of celebration—during festivals, New Year, good rains, or even when coffee prices increased. Hunting could take place in coffee plantations as well as in small patches of reserved and revenue forests adjacent to the plantations.

An owner of a medium-sized coffee plantation recalled his hunting experiences from earlier days:

> Those days [1980–86] we used to get good game around the estate itself. Along with my brothers and neighbours, I would go hunting at least once every fortnight. Those days we used to get plenty of sambar, chital and wild boar. But my share from the hunt is still sent to my home on all hunting occasions. You see, when you are part of a hunting group, you are 'in' even if you are absent. While the members get around two-and-a-half to three kg of meat, the hunting dogs get their share too ... these days, it is mostly wild boar meat; occasionally a sambar or barking deer, and once in a while a chital if you get really lucky.

He had moved from his plantation to live in Chikmagalur in 1986 and has not gone hunting since then, whereas many of his old-time hunting partners still continue to hunt. He also said that till date, his hunting group had never been caught by the Forest Department.

A woman planter asserted that there was no hunting in the region anymore. She said, 'Now the circumstances have changed and people fear the law.' Furthermore, she told us that the hunting of wild boar was very common, and that it was mostly for crop protection. She was not aware that the wild boar and hare were protected by the law and that it was illegal to hunt them. About 70 per cent of the planters interviewed said that wild boar do not damage coffee bushes, and 56 per cent of the planters were not aware about the existing legislations for protection of wildlife and the legal regulations for prevention of hunting.

According to a database maintained by a local wildlife conservation group, which is engaged in combating poaching in Chikmagalur, approximately 71 big and small hunting groups are active in the district. In order to highlight the severity of this issue, a representative from this organisation said:

> On an average, these groups organise hunts twice a month. For a 15-member hunting party where the share per head is 2.5 to 3 kg of meat, they need to kill at least one or two large animals [herbivores] in each hunt. By that estimate, the number of large animals being hunted every month could be as high as 70, or 850 for the whole year, in the district.

No wild animal listed in Schedules I–IV of the Wildlife Protection Act can be killed, irrespective of its location. Coffee plantations come under the purview of the Act. But hunting did occur in these private properties without the Forest Department's intervention. In the region, most of the hunting occurred outside the Bhadra Wildlife Sanctuary. All the respondents were aware of the risk and consequences of hunting in the Sanctuary. They also felt that hunting in reserved and revenue forests, as in coffee plantations, did not involve the same risk as hunting in the Sanctuary.

The number of hunting cases booked by the Forest Department, too, reflected this. According to their records, in a five-year period (2000–05), only 19 cases were booked for wildlife offences taking place outside the Bhadra Wildlife Sanctuary and only on four occasions were arrests made. During 2005, a single case was booked for the hunting of the sambar.

Enforcement of the Wildlife Protection Act by the Forest Department is seen as a major requirement. Although some planters have banned hunting in their estates, it is purely due to personal motivation and not very effective in combating hunting. Some planters suggested that such efforts would be more effective if there was support from the Forest Department. However, the Department did not consider enforcement outside the PA a priority. In fact, a senior officer of the Forest Department remarked:

> Bhadra Wildlife Sanctuary is full of animals to its capacity. Because of competition for space and resources, the old and weaker ones get booted out into the coffee plantations. If they cross their designated territory and enter human territory, they deserve to be killed. Policing in a planter's backyard is just not our job.

Another senior forest officer expressed their inability to patrol outside the Sanctuary due to a lack of resources. He said that there were insufficient guards to patrol the 492 sq. km area of the Bhadra Wildlife Sanctuary, and a lack of resources such as vehicles and arms. The Forest Department said they could act only if they were informed about any crime taking place outside the PA.

While expressing his support, a forest guard remarked: 'I go for beat with a *lathi* (stick). The poacher waits for me with a gun. What can I do to catch him?'

Shady Tales

We recorded 93 species of native trees in coffee plantations, characteristic of the dry and moist deciduous forests in the region (Bali et al. 2007). Tree-felling, timber-logging and shade-lopping of native trees occurred in the estates. The KPTA is intended to prevent any kind of damage to native trees. This includes cutting or felling, severing the trunk from the roots, uprooting the tree, burning, girdling or applying arboricides to a tree. However, all these activities took place openly as and when the need arose. Most plantation owners were not aware of the regulations regarding the felling of trees. Almost 50 per cent of the respondents believed that restrictions were limited to the commercial exploitation of native tree species. They believed that if they did not sell the timber, there was no need for permission from the Forest Department to cut a tree on their property. This misconception has had a major negative impact on the way shade is managed.

Shade management, an annual exercise of thinning of the shade, is carried out before the monsoon to ensure that coffee bushes receive sufficient sunlight during the season. Generally, it involves the lopping of branches of shade trees and occasionally, the removal of a few trees. It is a fairly common practice to cut a tree, irrespective of the species, without the Forest Department's permission. If required, trees are killed by girdling. In plantations that are at higher altitudes, intense shade-lopping takes place where all branches are cut leaving only the main stump. While the lopping of branches more than 10 centimetres in girth is not allowed, no one follows this regulation. In fact, 66 per cent of the respondents did not even know about this restriction.

There were other simpler ways of acquiring permits for timber harvest. One respondent elaborated on this for us:

> Silver oak is free. No permit required for its harvest. But we need a permit to transport it. Also, jungle wood harvest is a problem. It takes long time to get permits [for it]. For us it is difficult to deal with the Forest Department. So we just give contract to timber agents. These agents are very efficient at getting permits. They have got a fixed 'cut' for each level, starting from the top till the guards. They have got things under control.

The KPTA provides for an annual tree census to be conducted in all private properties as a means to keep a check on violations of any of the provisions of the Act. However, the estate managers reported that the tree census had never been conducted in the estates. One manager informed us that he was not aware about this census:

> If there is excess shade, we need to remove it. Timber from silver oak is sold in the market and jungle wood is used as fuel wood by our labourers. It is part of shade management. But we have no need to worry, as no officer will come to check in the estates.

A young planter, one of the few who was well-informed about the Act, pointed out that even though the KPTA provides for a decision on an application for timber harvest within a three-month period, he had been waiting for at least 18 months for the permission to harvest an old rosewood tree. However, the most interesting response came from an elderly owner of a very large plantation. He believed that the KPTA did not affect them: 'For us there is a totally different procedure for timber harvest ... one just needs to pay some bribe and in case of valuable trees like rosewood, give them their share of timber too.'

There was very low awareness among the planters about the direct implications of the Act on the management of coffee plantations. While all plantation owners interviewed were aware about at least the existence of 'some tree law', only around 8 per cent of the respondents had reasonable knowledge regarding this law. There was a lack of information about even the most relevant regulations of the Act. Furthermore, there was also a lack of implementation by the Forest Department. In the absence of the tree census, there was no way that the Forest Department could determine non-compliance of the law. Hence, even though it is the intention of the State to encourage the plantations of native trees and restrict felling, the KTPA seems to be inadequate by itself to preserve native trees in coffee plantations.

Good Policy, Bad Practice

Landscape ecology (Forman 1995; Hanski and Gilpin 1991) and more recently, reconciliation ecology approaches (Rosenzweig 2003) suggest that a range of secondary habitats such as agricultural lands should be included and managed in an integrated manner for effective biodiversity conservation. This can significantly influence the survival of species in remnant patches and increase the overall area available for conservation (Janzen 1998; Seibert 2002). Therefore, it is imperative that rather than only insisting on protecting habitats from human use, we also focus on finding ways to utilise the existing conservation potential of human-dominated landscapes that cover most of the terrestrial surface of the earth (Rosenzweig 2003). The primary focus of such efforts could be agricultural lands because a significant proportion of land worldwide is under agricultural use (UNEP 2001).

Coffee is the second-most traded commodity in the world (International Coffee Organization 2009). Studies from around the world suggest that traditionally managed coffee plantations can harbour a wide variety of biodiversity compared to other commercial crops because of the structural complexity of the habitat (Daily et al. 2003; Moguel and Toledo 1999; Perfecto et al. 1996; Vandermeer and Perfecto 2007). As a result, in the last decade, there has been considerable interest in traditional coffee systems. Many conservation organisations such as the Smithsonian Migratory Bird Center, Rainforest Alliance, Audobon Society and World Wildlife Fund are working towards the promotion of traditional practices in coffee cultivation as a means of biodiversity conservation in the tropics.

In the Western Ghats, the plantations are entirely shade-grown and to a large extent, traditionally managed. An abundance of native forest trees (for example, *Ficus spp.*) makes these plantations attractive to various organisms. During our sampling for mammals in coffee plantations, we recorded 28 species, characteristic of the dry and moist deciduous forest in the region, which included resident species as well as those spilling over from the adjoining forests (Bali et al. 2007). They are also used as corridors by several species (for example, elephants, *Elephas maximus*) for movement between patches of forests. Although the protection of natural habitats remains the backbone of biodiversity conservation strategies, these plantations could play an important supporting role (Anand et al. 2008; Bhagwat et al. 2008; Shahabuddin 1997; Shankar Raman 2006).

Imperfect Implementation

The long tradition of hunting has had a significant impact on wildlife in and around coffee plantations (Kumara and Singh 2004; Madhusudan and Karanth 2002). Hunting has been practised by landowners since colonial times (Elliot 1898) and has continued after Independence. Since its enactment in 1972, the Wildlife Protection Act has caused major changes in the way hunting is viewed in the country. It transformed what was a cultural rite amongst local communities into an illegal act in the eyes of the State. Since the two groups have different environmental concerns, this poses constraints in successful implementation of the law. Even though it is illegal to hunt a wild animal irrespective of its location, it is practically impossible to control hunting outside formally protected areas. Hence the practice of hunting has continued even after the Act was passed.

However, an interesting development is that the current generation of planters is getting educated in big cities. As a result, they are losing their hunting skills as well as their interest in hunting as a recreational activity. Some have even become ardent supporters of wildlife conservation and have formed local non-governmental organisations (NGOs) to monitor poaching of wildlife in the area.

Perverse Consequences

Coffee production methods have been through sweeping changes over the past few decades. There have been fluctuations in coffee prices in most coffee-producing countries, which have resulted in the technification of coffee cultivation, i.e., increased pesticide use, monoculture shade and drastic reduction in the number of shade trees, and the advent of more disease-resistant crops, leading to considerable loss of biodiversity (Greenberg et al. 1997; Perfecto and Snelling 1995; Vandermeer and Perfecto 2007).

Indian coffee accounts for only approximately 4.5 per cent of the world coffee production (DoC 2009). However, the price of coffee is determined by outside factors such as the global production and reference price. More than 80 per cent of coffee-growing units are owned by small landowners, who find it difficult to adjust to such market fluctuations. In the recent past, the performance of the coffee market in India has been so poor that most growers have not even been able to realise their investments (ibid.). Such unsteady market conditions lead the planters to supplement their income by harvesting timber on a large scale to make up for losses in the coffee market. Consequently, they have been planting more fast-growing timber species for securing future income. The large-scale conversion of diverse shade into monocultures of silver oak in the Western Ghats likely has negative effects on biodiversity (Bali et al. 2007; Damodaran 2002).

Ironically, due to the constraints imposed by the KPTA regulations on felling native trees, the private landowners plant more non-native timber species. Thus, a law that intended to protect native species is actually having the opposite effect and is leading to a decline in native species in human-dominated landscapes. Often, well-intentioned conservation policy has unintended consequences in a wide variety of contexts and cases, including leasehold forestry in Nepal (Baral and Thapa 2003), forest reserves in Latin America (Bobenrieth 1997), and fisheries in northern America and Europe (Degnbol and McCay 2006). Among other factors, uncertainty appears to play a key role in influencing local practices.

Policy and Practice: A Common Substrate

The protection of forests and wildlife protection is beset with handicaps, even when there is adequate policy and legislation. To begin with, law enforcement capabilities are plagued by insufficient infrastructure and resources. More serious problems in forest protection are, however, external to the organisation. This case demonstrates that formal policy and legislations by themselves do not necessarily result in appropriate conservation actions and practices at the local level. A formal policy or law can be a useful guiding tool, but cannot ensure conservation of wildlife outside PAs, nor of native trees in private lands. Social change is required to promote a wider conservation ethic in society, a transformation from the current practice towards the perceived ideal (Chapin et al. 2006), which cannot be achieved and sustained through only policing and top-down approaches. For instance, the increase in silver oak abundance is an unintended consequence of a policy that intends to protect native trees in private landscapes. This is because different stakeholders (for example, the State Forest Department and coffee planters) have different priorities. Hence, such laws, being so broadly applied and ineffectively implemented, are virtually meaningless in practice (Figure 9.2).

Figure 9.2: Conceptual Model of the Social–Ecological System that Exists Outside PAs and the Role of Institutions

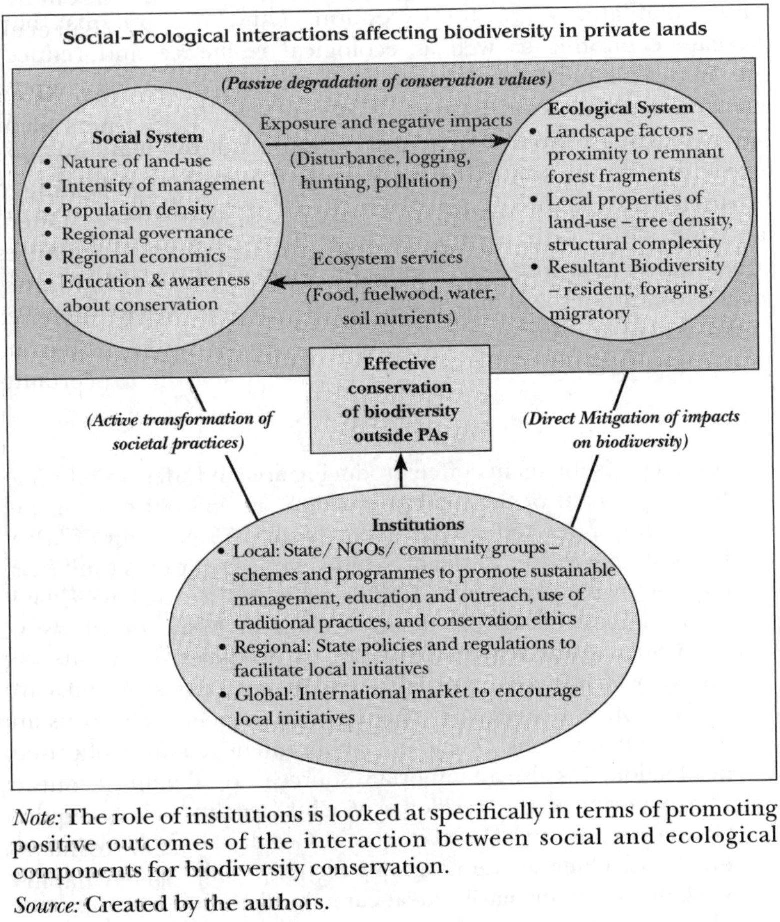

Note: The role of institutions is looked at specifically in terms of promoting positive outcomes of the interaction between social and ecological components for biodiversity conservation.

Source: Created by the authors.

We therefore suggest that successful and sustainable 'transformations' need a more synergistic approach i.e., searching for common grounds and integration of the objectives of various stakeholders—a shift in policies from being top-down directives to a more inclusive approach that integrates community development as an inherent goal and encourages community participation (Figure 9.2). For instance, the support for biodiversity conservation and the retention of diverse native shade trees can create better marketing opportunities for coffee planters. Thus, by helping create such opportunities, the State can make wildlife conservation in these privately owned landscapes pay for itself.

All nature–society interactions have trade-offs and these coffee plantations are no exception. Land use options that increase economic as well as ecological resilience and reduce the vulnerability of local communities and natural ecosystems are fundamental to conservation efforts that focus on human landscapes since biodiversity conservation is not the main goal of agriculture. In the context of the Western Ghats, there is a need to broaden conservation efforts to be inclusive of the matrix of human land use along with natural habitats. This calls for rethinking conservation planning and design approaches that are inclusive of local communities and aim at livelihood improvement as a means to the goal of conservation objectives.

Notes

1. Karnataka is the main coffee producing area in India, accounting for 71 per cent of the total production, and 58 per cent of the total land under coffee cultivation. Around 75 per cent of large landholdings fall in Karnataka. Over 43 per cent of Karnataka's coffee-growing area lies in Chikmagalur district, making it one of the largest coffee producing regions in India. In addition, the Chikmagalur region is the largest producer of the *arabica* variety, which needs approximately 70 per cent shade. All of Indian coffee is essentially shade-grown, hand-picked and sun-dried on plantations. Due to the labour-intensive nature of coffee production, it is also an important source of rural employment.
2. The majority of the residents of Chikmagalur belong to the Gowda community. The two sects of Gowdas are Vokkaliga and Lingayats. Whereas the Lingayats are essentially vegetarian, the Vokkaligas are primarily meat-eaters and skilled hunters with

a long tradition of hunting and meat-sharing. The Gowdas constitute around 60 per cent of the population of planters in the region. The Kodava (the main ethnic group in the Coorg region), Muslims and South Canara Christians constitute most of the remaining 40 per cent. All of these are predominantly non-vegetarian communities. The coffee labourers mainly fall into three categories—residents, non-residents and migratory. Most labourers are residents in the estates. They have higher wages than the other two categories of labourers, and are provided accommodation and fuelwood. These are mainly the Tulus from South Canara and the Tamils from the neighbouring state, Tamil Nadu, who moved into the coffee-producing regions during the post-Independence (1947) period in large numbers to find employment. The non-resident labourers return everyday to work from nearby villages and small settlements. The migratory labourers are seasonal workers, who come in during the coffee-picking season for a month or two and reside in the estates. They are mostly the Lambanis from North Karnataka or small farmers from surrounding villages who come looking for work after harvesting their own fields. The Lambanis receive the lowest wages. In general, coffee labourers get lower wages than most other daily-waged jobs such as mining and road construction. They do not have rights to the coffee plants or estate trees but are often provided fuelwood by the estate owners.

3. A preliminary round of interviews with the key informants revealed a negative impact on wildlife due to widespread hunting, and large-scale conversion of diverse, natural shade to monoculture shade of non-native timber species such as silver oak (*Grevillea robusta*), which provide an additional source of income. A review of the literature highlighted certain policies and legislations that intend to facilitate the conservation of ecological values of the human-impacted habitats. Two main legal instruments—the KPTA, 1976 and Indian Wildlife Protection Act, 1972—were identified as being most relevant to conservation. We analysed the two legislations with regard to their content and lacunae in this regional context. We also attempted to examine the effectiveness of implementation of these acts by the Forest Department and the challenges faced by them. We examined the perceptions of coffee planters towards conservation and their awareness regarding the two acts.

Data were collected from December 2005 to June 2006, through structured interviews, group discussions and direct observations.

In total, 65 extensive interviews were conducted. Questions to the coffee planters were focused around five general themes (Appendix 9.1): (i) description of plantation management practices and general information about the family/group; (ii) practices relating to the conservation of native trees in plantations; (iii) practices relating to the conservation of wildlife in plantations; (iv) information about various laws and State policies and awareness of conservation issues; and (v) general attitude towards participation in conservation of wildlife in the region. The State Forest Department representatives were asked questions related to the implementation of these two Acts. In addition to these structured interviews, several informal discussions were conducted opportunistically to get an understanding of the landscape and culture. We have used the data qualitatively as a case study to get at the questions stated above. We triangulated information from multiple sources to ensure and verify the validity of the data. We attempted to get an equal representation of different groups among the respondents—planters, labourers, officers and staff of the State Forest Department as well as people from non-plantation backgrounds, in order to get better insights into these issues (Table 9.2).

Table 9.2: Respondents Interviewed During the Study

Interviewees and Questionnaire Respondents	*Number*
Plantation representatives:	
a) Owners and managers	24
b) Labourers	20
c) Coffee growers' associations	3
Coffee Board of India	2
Karnataka Forest Department	5
Representatives of local wildlife conservation NGOs	3
Miscellaneous professionals outside the coffee planters' community —Academicians (2), journalists (2), lawyer (1), auditor (1), government land surveyor (1), wildlife photographer (1)	8

Source: Field studies by the authors.

References

Anand, M. O., J. Krishnaswamy, and A. Das. 2008. 'Proximity to Forests Drives Bird Conservation Value of Coffee Plantations: Implications for Certification', *Ecological Applications* 18(7): 1754–63.

Bali, A., A. Kumar, and J. Krishnaswamy. 2007. 'The Mammalian Communities in Coffee Plantations around a Protected Area in the Western Ghats, India', *Biological Conservation* 139(1–2): 93–102.

Baral, J. C., and Y. B. Thapa. 2003. 'Nepal's Leasehold Forestry for the Poor: Looking at the Unintended Consequences', Mountain Forum Library Online Library and Reference Database. Available at http://www.mtnforum.org/resources/library/barax03b.htm (accessed 1 May 2006).

Bhagwat, S. A., K. J. Willis, H. J. B. Birks, et al. 2008. 'Agroforestry: A Refuge for Tropical Biodiversity?', *Trends in Ecology and Evolution* 23(5): 261–67.

Bobenrieth, E. S. A. 1997. 'Forest Reserves in Latin America: Environmental Conflict, Regulation Uncertainty, and Forest Stock Response'. In Alan Richards (ed.), *The Political Economy of International Environmental Cooperation*, pp. 19–26. San Diego: Institute on Global Conflict and Cooperation.

Brockington, D. 2001. *Fortress Conservation: The Preservation of the Mkomazi Game Reserve, Tanzania.* Oxford: James Currey.

Chapin, F. S., III, M. W. Oswood, K. Van Cleve, et al. (eds). 2006. *Alaska's Changing Boreal Forest.* New York: Oxford University Press.

Daily, G. C., G. Ceballos, J. Pacheco, et al. 2003. 'Countryside Biogeography of Neotropical Mammals: Conservation Opportunities in Agricultural Landscapes of Costa Rica', *Conservation Biology* 17(6): 1814–26.

Damodaran, A. 2002. 'Conflict of Trade-Facilitating Environmental Regulations with Biodiversity Concerns: The Case of Coffee-Farming Units in India', *World Development* 30(7): 1123–35.

Das, A., J. Krishnaswamy, K. S. Bawa, et al. 2006. 'Prioritisation of Conservation Areas in the Western Ghats, India', *Biological Conservation* 133(1): 16–31.

Degnbol, P., and B. J. McCay. 2006. 'Unintended and Perverse Consequences of Ignoring Linkages in Fisheries Systems', *ICES Journal of Marine Science* 64(4): 793–97.

Department of Commerce (DoC). 2009. 'Rangachari Report on Plantations'. Ministry of Commerce and Industry, New Delhi, Government of India.

Dutta, R., R. K. Singh, and A. Mookerjee. 2004. *Wildlife Law: A Ready Reckoner*, New Delhi: Wildlife Trust of India.

Elliot, H. R. 1898. *Gold, Sport, and Coffee Planting in Mysore.* Westminster, UK: Archibald Constable & Co.

Food and Agricultural Organization (FAO). 2006. *Global Forest Resources Assessment 2005: Progress Towards Sustainable Forest Management.* Rome, Italy: Food and Agriculture Organisation, United Nations.

Forman, R. T. T. 1995. *Land Mosaics: The Ecology of Landscape and Regions.* Cambridge, UK: Cambridge University Press.

Gardner, T. A., J. Barlow, R. Chazdon, et al. 2009. 'Prospects for Tropical Forest Biodiversity in a Human-Modified World', *Ecology Letters* 12(6): 561–82.

Greenberg, R., P. Bichier, A. C. Angon et al. 1997. 'Bird Populations in Shade and Sun Coffee Plantations in Central Guatemala', *Conservation Biology* 11(2): 448–59.

Hanski, I. A., and M. E. Gilpin. 1991. 'Metapopulation Dynamics: Brief History and Conceptual Domain', *Biological Journal of the Linnaean Society* 42(1–2): 3–16.

International Coffee Organization. 2009. 'About Us'. Available at http://www.ico.org/ (accessed 28 May 2009).

Janzen, D. 1998. 'Gardenification of Wildland Nature and the Human Footprint', *Science* 279(5355): 1312–13.

Karanth, K. K., V. Sankararaman, S. Dalvi, et al. 2016. 'Producing Diversity: Agroforests Sustain Avian Richness and Abundance in India's Western Ghats', *Frontiers in Ecology and Evolution* 4: 111.

KFD. 2000. 'Karnataka Preservation of Trees Act 1976, as Amended by the Amending Act 20 of 2000'. Available at http://dpal.kar.nic.in/pdf_files/76%20of%201976%20%28E%29.pdf (accessed 18 May 2007).

Kumara, H. N., and M. Singh. 2004. 'The Influence of Differing Hunting Practices on the Relative Abundance of Mammals in Two Rainforest Areas of the Western Ghats, India', *Oryx* 38(3): 321–27.

Madhusudan, M. D., and K. U. Karanth. 2002. 'Local Hunting and the Conservation of Large Mammals in India', *Ambio* 31(1): 49–54.

McNeely, J. A., and G. Schroth. 2006. 'Agroforestry and Biodiversity Conservation—Traditional Practices, Present Dynamics, and Lessons for the Future', *Biodiversity and Conservation* 15(2): 549–54.

Menon, S., and K. S. Bawa. 1997. 'Applications of Geographical Information System, Remote Sensing and a Landscape Ecology Approach to Biodiversity Conservation in Western Ghats', *Current Science* 73(2): 134–45.

MoEF. 2009. 'National Forest Policy 1988'. New Delhi, Government of India. Available at http://envfor.nic.in/legis/forest/forest1.html (accessed 8 April 2008).

Moguel, P., and V. M. Toledo. 1999. 'Biodiversity Conservation in Traditional Coffee Systems in Mexico', *Conservation Biology* 13(1): 11–21.

Myers, N., R. A. Mittermeler, C. G. Mittermier, et al. 2000. 'Biodiversity Hotspots for Conservation Priorities', *Nature* 403(6772): 853–58.

Perfecto, I., and R. Snelling. 1995. 'Biodiversity and the Transformation of a Tropical Agroecosystem: Ants in Coffee Plantations', *Ecological Applications* 5(4): 1084–97.

Perfecto, I., R. A. Rice, R. Greenberg, et al. 1996. 'Shade Coffee: A Disappearing Refuge for Biodiversity', *Bioscience* 46(8): 598–99.

Rangarajan, M. 2001. *India's Wildlife History: An Introduction.* New Delhi: Permanent Black.

Rodrigues, A., S. J. Andelman, M. I. Bakarr, et al. 2004. 'Effectiveness of the Global Protected Area Network in Representing Species Diversity', *Nature* 428(6983): 640–43.

Rosenzweig, M. L. 2001. 'Loss of Speciation Rate will Impoverish Future Diversity', *Proceedings of the National Academy of Sciences* 98(10): 5404–10.

———. 2003. 'Reconciliation Ecology and the Future of Species Diversity', *Oryx* 37(2): 194–205.

Shahabuddin, G. 1997. 'Preliminary Observations on the Role of Coffee Plantations as Avifaunal Refuges in the Palani Hills of Western Ghats', *Journal of Bombay Natural History Society* 94(1): 10–21.

Shankar Raman, T. R. 2006. 'Effects of Habitat Structure and Adjacent Habitats on Birds in Tropical Rainforest Fragments and Shaded Plantations in the Western Ghats, India', *Biodiversity and Conservation* 15(4): 1577–1607.

Siebert, S. F. 2002. 'From Shade- to Sun-Grown Perennial Crops in Sulawesi, Indonesia: Implications for Biodiversity Conservation and Soil Fertility', *Biodiversity and Conservation* 11(11): 1889–1902.

UNEP. 2001. 'Global Biodiversity Outlook: Status and Trends in Global Biodiversity', Secretariat of the Convention on Biological Diversity (CBD), Montreal, Canada. Available at http://www.cbd.int/gbo1/chap-01-07.shtml (accessed 8 January 2009).

Vandermeer, J., and I. Perfecto. 2007. 'The Agricultural Matrix and a Future Paradigm for Conservation', *Conservation Biology* 21(1): 274–77.

Western, D. 2001. Human-Modified Ecosystems and Future Evolution. *Proceedings of the National Academy of Sciences* 98(10): 5458–65.

Appendix 9.1:
The Wildlife Protection Act of India, 1972

In 1972, the Government of India enacted a comprehensive legislation, the Wildlife Protection Act, with the objective of effectively controlling poaching and illegal trade in wildlife and its derivatives. The Act is a criminal law that provides ways to deal with offences related to wildlife. It is also a land acquisition law, instituted to facilitate the creation of PAs as a means to conserve biodiversity. At the same time, it is a natural resources management law, controlling land and resource usage pattern in the PAs, which relates, in turn, to the conservation and protection of endangered species (Dutta et al. 2004).

The Act provides for the protection of wild animals, birds and plants, and matters related to them. It extends to the whole of India, except the state of Jammu and Kashmir. The definition of 'wildlife' is very broad, extending to both flora and fauna, which includes any animal, bees, butterflies, moths, crustaceans and fish, and aquatic or land vegetation which form part of any habitat. The term 'animal' includes amphibians, birds, mammals, reptiles and their young, and in the case of birds and reptiles, their eggs. As per the provisions of the Act, all wildlife is protected in national parks and sanctuaries.

There are six schedules appended to the text of the Act, containing a list of species protected by it. Schedules I and II include most of the endangered and threatened animals of India. 'Wild animal' is any animal that is (i) listed in Schedule I– IV of the Act and (ii) is found in nature.

The Act prohibits the hunting of any wild animal specified in Schedules I, II, III and IV. Only in the case of an animal becoming dangerous to human life or property, being disabled or diseased beyond recovery, or in self-defence, hunting is not considered an offence. By definition, 'hunting' includes: (a) capturing, killing, poisoning, snaring and trapping of any wild animal, and every attempt to do so; and (b) injuring or destroying or taking any part of the body of any such animal, or in the case of wild birds or reptiles, damaging the eggs of such birds or reptiles, or disturbing their eggs or nests. Also, 'trophy' means the whole or part of any wild animal, other than vermin, which has been kept or preserved by any means, such as skins, antler, horn, feather, nail, tooth, etc.

Any person who disregards the provisions of this Act is considered guilty and can be punished with up to 3 years of imprisonment, a fine of up to Rs 25,000, or both. Also, the court may direct for the cancellation of arms license granted to such person under the Arms Act, 1959 for the possession of an arm with which an offence against this Act is committed.

Appendix 9.2: The Karnataka Preservation of Trees Act (KPTA), 1976

The KPTA is defined as an act to make better provision for the preservation of trees in the state. This Act is intended to ensure the preservation of trees in the state by regulating the felling of trees and tree-planting towards ecological balance. It came into force on 17 July 1976.

'Preservation of trees' includes maintenance and protection of trees to encourage normal growth and prevention of damage or cutting or felling. 'To fell a tree' means severing the trunk from the roots, uprooting the tree, and it includes burning, cutting, girdling or applying arboricides to a tree to cause substantial damage or destruction to it. There is provision for the appointment of a 'tree officer' by the Principal Chief Conservator of Forests (PCCF) and a 'tree authority' consisting of five members for each urban and rural area. Duties of the 'tree authority', among others, are: to preserve all trees within its jurisdiction (sec. 7[a]); carry out census of the existing trees (sec. 7[b]); and to specify the kinds of trees which each locality and type of land should have (sec. 7[c]).

There are various restrictions on the felling of trees. No person can fell a tree, whether in his ownership or occupancy, without prior permission of the tree officer (sec. 8). The permission can be granted in specific cases such as, if the tree:

i) is dead, diseased or wind-fallen;
ii) has silviculturally matured;
iii) constitutes a danger to life or property;
iv) constitutes obstruction to traffic;
v) is substantially damaged or destroyed by fire, lightning, rain or other natural causes; and
vi) is required to be removed for cultivation (including coffee cultivation) or bona-fide use (sec. 8[3] [vi] inserted in 1998).

The tree officer is required to make all possible endeavours to dispose of any application received within three months' time (sec. 8, sub-sec. 4 inserted in 1998). However, the restrictions on felling a tree are not applicable to certain species (plantation and exotic trees such as coconut, rubber, silver oak and eucalyptus), including silver oak (sub-sec. 7 inserted in 1977 and 1987). Number of trees planted in coffee plantations as specified by notification 1875 is a minimum of 750 trees per hectare.

Appendix 9.3: Set of Questions for Coffee Planters (Based around Five Broad Themes)

Respondents: Planters

a) Plantation owners: covered all categories (individual and company holders; small and large holders)
b) Plantation labourer
c) Associations of coffee planters (representatives from Karnataka Planters' Association [KPA], United Planters Association of Southern India [UPASI])
d) Coffee Board of India

I. Background Information, Coffee Management Practices

- Name
- Size of the coffee estate
- Type of holding (private/company)
- Location of the estate (proximity to forests)
- Number of years in coffee plantation
- Type of coffee: *arabica* or *robusta*, polyculture or monoculture
- Fencing of plantations
- Use of pesticides
- Price of coffee in national and international markets
- Costs of inputs into coffee production
- Production, profit–loss margin
- Subsidies/rebates that they are enjoying
- Side business, if any (for example, coffee tourism)

II. Practices Relating to Conservation of Native Trees in Plantations

- Approximate proportion of silver oak trees in their estate
- Frequencies of timber harvest
- What is the procedure for timber harvest from silver oak?
- What is the procedure for timber harvest from native species?
- What tree species do they replant (in place of harvested trees)?
- Reasons for planting so much silver oak
- Economic returns from silver oak
- Effects of silver oak on coffee plants or soil, if any
- Views on 'tree conservation'

III. Practices Relating to Conservation of Wildlife in Plantations

- What wildlife occurs in their plantation?
- Problems faced due to wild animals in the estates, if any:
    - Attacks on humans

  - Attacks on domestic animals
  - Damage done to coffee crop
- How do they deal with the wildlife-related problems?
  - Do they seek help from the Forest Department?
- Benefits derived from wildlife in plantations, if any:
  - Utilitarian as well as non-utilitarian benefits
- Do they hunt? If yes,
  - How often, with whom (alone/groups), what species?
  - Describe a typical hunt.
  - Do they hunt in their estates?
  - Or do they hunt in the forest (sanctuary/reserved forest)?
- Do the labourers hunt in the estate?
- What methods are used (guns/snares/others)?
- What is the attitude of planters towards hunting by labourers? Oppose/ignore/encourage it
- What changes have occurred in animal diversity/abundance over last few decades? (declines)
- What are the reasons behind this?
- Views on 'wildlife conservation'

IV. Information Regarding Existing Policies and Laws

- Are there any laws to protect trees?
- Have they ever received any notice from the Forest Department for non-compliance with any regulation?
- Have they ever sought help/permissions from the Forest Department in any matter related with trees?
- Are they aware about the regulations for shade-lopping (cannot cut a branch greater than 10 centrimetres girth)
- What do they think about this sort of regulation? Absurd/justified
- What are the reserved/protected trees? (List them)
- Is this law useful/serving the purpose?
- Are there any laws to protect wild animals?
- If yes, what animals are protected? (List them)
- If they hunt—have they ever got into trouble with the Forest Department?
- Do they know of anyone who has been caught in relation with hunting matters?
- What does the Forest Department do to protect wildlife outside the sanctuary?
- What is the intensity of hunting in the region (district)?
- What is the way forward? (Ask for their suggestions)
- Is the Wildlife Protection Act a good/comprehensive/useful law?

V. General Attitude/Awareness Level
(Future of Conservation Values of Coffee Landscapes)

- What do they think about the liberalisation of the coffee market/ free market (what were the ups and downs in last 10 years)?
- What are the traditional ways of coffee cultivation?
- What are the present cultivation practices?
- What has changed? Why?
    - Better pest management?
    - Increased market pressure?
    - Local/regional climate?
    - Others?
- Attitudes towards conservation (economic or ecological)
    - Understanding about ecosystem services
    - Ecological significance of biodiversity
- What is sustainable coffee cultivation?
- What are the perceived economic benefits (if any) of being close to forest landscapes?
- What do they think about the 'bird-friendly' coffee programmes being run in other parts of the world?
- To shift from non-eco-friendly to eco-friendly practices:
    - What could be the change in production outputs/economic returns?
    - How much compensation do they expect?
    - What could be the more direct incentives of traditional practices rather than the option of compensation?

Respondents: Forest Department

- How are the Wildlife Protection Act and KPTA applied in private lands?
- How are various regulations of the two Acts implemented outside of PAs?
- What are the hindrances to law enforcement in private lands?
- What is the scope of community participation in wildlife and native tree conservation around PAs?

Bibliography

Andam, K. S., P. J. Ferraro, A. Pfaff, et al. 2008. 'Measuring the Effectiveness of Protected Area Networks in Reducing Deforestation', *Proceedings of the National Academy of Sciences* 105(42): 16089–94.

Anonymous. 2003. 'Wildlife on the Run', *The Hindu*, Magazine, online edition, 22 June 2003. Available at http://www.hinduonnet.com/mag/2003/06/22/stories/2003062200140700.htm (accessed 8 August 2008).

———. 2008. 'New Coastal Management Plan Jeopardises Livelihood of Fishermen', *The Hindu*, 10 June 2008.

Bose, A., C. Garcia, and B. Vira. 2019. 'Mottled Motivations and Narrow Incentives: Exploring Limitations of Direct Incentive Policies in the Western Ghats, India', *Ecological Economics* 156: 511–18.

Buscher, B., R. Fletcher, D. Brockington, et al. 2017. 'Half-Earth or Whole Earth: Radical Ideas for Conservation, and their Implications', *Oryx* 51(3): 407–10.

Daniels, R. J. R. 2003. 'Impact of Tea Cultivation on Anurans in the Western Ghats', *Current Science* 85(10): 1415–22.

Dupont, Veronique. 2004. 'Socio-Spatial Differentiation and Residential Segregation in Delhi: A Question of Scale?', *Geoforum* 35(2): 157–75.

Forest Department, Government of NCT of Delhi. Available at http://delhi.gov.in/wps/portal/!ut/p/c0/04_SB8K8xLLM9MSSzPy8xBz9CP0os3hvdxMXExdTEwMDXzNzA09_YyOPEHcnAwMLQ_2CbEdFAIa72s0!/?WCM_PORTLET=PC_7_KG4D4D5400M670IO32HTGB0046_WCM&WCM_GLOBAL_CONTEXT=/wps/wcm/connect/doit_forest/Forest/Home/Delhi+State (accessed on 12 April 2017).

Gadgil, M., and R. Guha. 1993. *This Fissured Land: An Ecological History of India.* California, USA: University of California Press.

Gaveau, D. L. A., H. Wandono, and F. Setiabudi. 2007. 'Three Decades of Deforestation in Southwest Sumatra: Have Protected Areas Halted Forest Loss and Logging, and Promoted Regrowth?', *Biological Conservation* 134(4): 495.

Ghertner, D. Asher. 2011. *Rule by Aesthetics: World-Class City Making in Delhi.* Oxford: Wiley-Blackwell.

Gururani, Shubhra. 2013. 'Flexible Planning: The Making of India's

"Millennium City", Gurgaon'. In Anne Rademacher and K. Sivaramakrishnan (eds), *Ecologies of Urbanism in India: Metropolitan Civility and Sustainability*, p. 304. Hong Kong: Hong Kong University Press.

Hoekstra, J. M., T. M. Boucher, T. H. Ricketts, et al. 2005. 'Confronting a Biome Crisis: Global Disparities of Habitat Loss and Protection', *Ecology Letters* 8(1): 23–29.

Johnsingh, A. J. T. 2006. *Field Days: A Naturalist's Journey through South and Southeast Asia.* Hyderabad: Universities Press.

Kapoor, K. S., S. Kumar, A. Rajasekaran, and R. S. Rawat. 2006. 'Chilgoza (Pinus gerardiana Wall): A Unique Conifer of Inner Dry Valleys of North West Himalaya'. In in S.S. Negi, R. K. Srivastava, and S. Nautiyal (eds.), *Studies on Himalayan Pines*, pp. 158–70. Dehradun: USDA, Forest Research Institute.

Khera, N., V. Mehta, and B. C. Sabata. 2009. 'Interrelationship of Birds and Habitat Features in Urban Greenspaces in Delhi, India', *Urban Forestry and Urban Greening* 8(3): 187–96.

Khoshoo, T. N. 1995. 'Against Cheetah'. *The Times of India*, 14 July 1995, p. 12.

Layton, Robert, and Peter J. Ucko (eds). 2003. *The Archaeology and Anthropology of Landscape: Shaping Your Landscape.* London and New York: Routledge.

Nagendra, H. 2008. 'Do Parks Work? Impact of Protected Areas on Land Cover Clearing', *Ambio: A Journal of the Human Environment* 37(5): 330–37.

Ninan, K. N., and J. Sathyapalan. 2005. 'The Economics of Biodiversity Conservation: A Study of a Coffee Growing Region in the Western Ghats of India', *Ecological Economics* 55(1): 61–72.

Rai, N. D. 2007. 'The Ecology of Income: Can We Have Both Fruit and the Forest?' In G. Shahabuddin and M. Rangarajan (eds), *Making Conservation Work*, pp. 142–62. Ranikhet: Permanent Black.

Rajasekaran, A., K. D. Sharma, and J. Singh. 2006. 'Natural Regeneration of Pinus gerardiana Wall in Himachal Pradesh', Simla: Himachal Forest Research Institute.

Rangarajan, M. 2001. *India's Wildlife History: An Introduction.* New Delhi: Permanent Black and Ranthambhore Foundation.

Sachs, J. D., and A. M. Warner. 1995. *Natural Resource Abundance and Economic Growth.* Cambridge, MA: National Bureau of Economic Research.

Saxena, A. K., and J. S. Singh. 1984. 'Tree Population Structure of Certain Himalayan Forest Associations and Implications Concerning their Future Composition', *Vegetatio* 58: 61–69.

Sheikh, Shahana, and Ben Mandelkern. 2015. 'Limbo in Sangam Vihar: Delhi's Largest Agglomeration of Unauthorised Colonies', A Report of the Cities of Delhi Project. New Delhi: Centre for Policy Research.

Singh, Vijai Shanker, Deep Narayan Pandey, and Neha Pandey Prakash. 2011. 'What Determines the Success of Joint Forest Management? Science-based Lessons on Sustainable Governance of Forests in India', *Resources, Conservation and Recycling* 56(1): 126–33.

Terborgh, J. 2010. 'Why We Must Bring Back the Wolf', *The New York Review of Books*, 15 July 2010.

Thapar, V. 2003. 'The Nineteenth Century—The Beginning'. In V. Thapar (ed.), *Battling for Survival: India's Wilderness over Two Centuries*, pp. 1–30. New Delhi: Oxford University Press.

Ticktin, T. 2004. 'The Ecological Implications of Harvesting Non-Timber Forest Products', *Journal of Applied Ecology* 41: 11–21.

Wilson, E. O. 2016. *Half-Earth: Our Planet's Fight for Life*. London: Liveright Publishing.

Notes on the Editors and Contributors

Meghna Agarwala is an ecologist who works on forest degradation and regeneration in central India. She is currently an Assistant Professor of Environmental Studies at Ashoka University, Sonipat, Haryana.

Ambika Aiyadurai is an anthropologist with a special interest in human–animal relations and biodiversity conservation. She is an Assistant Professor (Anthropology) in the Department of Humanities and Social Sciences, Indian Institute of Technology (IIT), Gandhinagar. Her long-term research aims to understand how local and global forces shape human–animal relations. As an SSRC Fellow, she was also part of the Global Asia Initiative at Duke University in 2018.

Archana Bali completed her Masters' degree in Wildlife Conservation from the National Centre for Biological Sciences in 2006, where she worked on mammals on coffee estates in southern Karnataka. She also carried out research on policy and practice relating to the impact of conservation laws outside protected areas. She then joined the University of Alaska, Fairbanks, as the first George Schaller Fellow for a PhD on human-caribou systems in the context of climate change. Archana passed away in 2014. Her PhD was awarded posthumously.

Ruth DeFries is an environmental geographer who specialises in the use of remote sensing to study Earth's habitability under the influence of human activities, such as deforestation, that influence regulating biophysical and biogeochemical processes. She is a recipient of the Macarthur Genius Award from the MacArthur Fellows Program (2007), and a member of the National Academy of Sciences, USA. She is currently Professor of Ecology and Sustainable Development at Columbia University.

T. Ganesh is a Senior Fellow at the Suri Sehgal Centre for Biodiversity and Conservation, Ashoka Trust for Research in Ecology and the Environment (ATREE), Bengaluru. His research interests lie in plant–animal interactions, tropical seed dispersal/ seed predation systems, long-term monitoring of phenology and vegetation dynamics of tropical rainforests. His current focus is on understanding the ecology and conservation of migratory grassland birds and grasslands in general.

Rajkamal Goswami is currently handling the research, monitoring and coordination component of a large-scale multi-institutional project on the mapping and inventory of bioresources in the north-eastern region of India based at Ashoka Trust for Research in Ecology and Environment (ATREE), Bengaluru. He is a National Geographic Society Explorer and is a recipient of the National Post-Doctoral Fellowship award by the Science, Engineering and Research Board, Government of India.

Y. V. Jhala is a wildlife biologist with an active research programme on Indian carnivores. He has supervised research projects on the Indian wolf, striped hyaena, golden jackal, Indian fox, Asiatic lion and tiger, using satellite and recapture telemetry. He is currently a senior scientist at the Wildlife Institute of India, Dehradun, a research associate at the Smithsonian Institution, Washington, D.C., and a member of the IUCN SSC Specialist Group for Wolves and Canids.

Kanchi Kohli is a researcher and writer working on environment, forest and biodiversity governance in India. Her work explores the links between law, industrialisation and environmental justice. Other than her independent work, Kanchi is presently Legal Advisor to the Centre for Policy Research (CPR)-Namati Environmental Justice Programme. Since 2004, she has co-coordinated an information dissemination service for forest and wildlife cases in the Supreme Court of India as well as the Campaign for Conservation and Community Control over Biodiversity.

Manju Menon is a senior fellow at Centre for Policy Research and Director of its Environmental Justice Program, implemented in collaboration with Namati, an international NGO dedicated to legal empowerment. Manju researches and writes on environmental law and policy and the regulation of infrastructure development. She investigates the political economy of environmental regulation in the context of energy projects, ports, mining and urban development.

Qamar Qureshi is involved with projects dealing with biodiversity assessment at the landscape and species levels and studying habitat use patterns using remote sensing and GIS, and his specialisations are in conservation biology, quantitative ecology and landscape ecology. He is currently a senior scientist at the Wildlife Institute of India, Dehradun.

Rinki Sarkar is an independent researcher based in Delhi working on development concerns and environmental issues across a wide geographical expanse of the mid-altitudinal Western Himalayas. Trained as an economist, her research studies have evolved to assume an interdisciplinary thrust. Rinki is a visiting faculty in the Environmental Studies Program of Ashoka University, Sonipat, Haryana. She also conducts rural field study workshops in the Western Himalayas for students of IIT Mandi and the Worcester Polytechnic, USA, and lectures on urban history at the School of Planning and Architecture, New Delhi.

Ghazala Shahabuddin is a Senior Scientist with the non-profit Centre for Ecology, Development and Research, India (CEDAR) and was earlier a Professor at Ambedkar University, Delhi (2009–14). Her research interests are in biodiversity concerns at the interface of forests and people. Her current research is on the patterns of extinction and adaptation in native forest bird communities in response to land use change in the Himalayan forests of Uttarakhand. Her publications include *Making Conservation Work* (Permanent Black, 2007), *Conservation at the Crossroads* (Permanent Black, 2010) and *Nature without Borders* (Orient BlackSwan, 2014).

Kartik Shanker is an ecologist with interests in inter-disciplinary research as well as conservation action and outreach. He is faculty at the Indian Institute of Science (IISc), Bengaluru, where he has worked on various aspects of the distribution and diversity of flora and fauna in terrestrial and marine ecosystems. He has also served as Director of the ATREE, Bengaluru, and is a founder-trustee of Dakshin Foundation, an NGO which works largely on coastal and marine conservation and development.

Neha Sinha works on environmental policy with the Bombay Natural History Society (BNHS), Mumbai. Currently she is working on Important Bird Areas (IBAs), a community conservation programme in Wokha, Nagaland, and a conservation project in Pench, Madhya Pradesh. Neha is also a columnist, writing on wildlife and environment for *The Hindu, The Wire, DailyO, Deccan Herald* and other newspapers. She is a reviewer for the *Economic and Political Weekly* and *Current Science* journals, and a Contributing Editor for *The Wire.*

K. Sivaramakrishnan is Dinakar Singh Professor of Anthropology and Professor of Forestry and Environmental Studies at Yale University, where he is also Co-Director of the Program in Agrarian Studies, and Co-Director of the Inter Asia Initiative. His published and continuing research deals with the colonial and contemporary history of forestry and wildlife in India, rural development, circular migration, environmental law, as well as urban ecology in India and Asia.

M. Vikas is a PhD student at the Nelson Institute for Environmental Studies, University of Wisconsin-Madison. He is currently working with adviser Paul Robbins on a project that examines the long-term socioeconomic impacts of afforestation in Himachal Pradesh, and also working on a PhD dissertation on human-leopard interactions in north-western India.

Index